End-to-End
Quality of Service
over Cellular Networks

End-to-End Quality of Service over Cellular Networks

Data Services Performance and Optimization in 2G/3G

Edited by

G. Gómez and R. Sánchez

Both of
Optimi Corporation
Spain

John Wiley & Sons, Ltd

Other Wiley Editorial Offices

John Wiley & Sons Inc., 111 River Street, Hoboken, NJ 07030, USA

Jossey-Bass, 989 Market Street, San Francisco, CA 94103-1741, USA

Wiley-VCH Verlag GmbH, Boschstr. 12, D-69469 Weinheim, Germany

John Wiley & Sons Australia Ltd, 33 Park Road, Milton, Queensland 4064, Australia

John Wiley & Sons (Asia) Pte Ltd, 2 Clementi Loop #02-01, Jin Xing Distripark, Singapore 129809

John Wiley & Sons Canada Ltd, 22 Worcester Road, Etobicoke, Ontario, Canada M9W 1L1

Wiley also publishes its books in a variety of electronic formats. Some content that appears in print may not be
available in electronic books.

British Library Cataloguing in Publication Data

A catalogue record for this book is available from the British Library

ISBN-13 978-0-470-01180-5 (HB)
ISBN-10 0-470-01180-7 (HB)

Typeset in 10/12pt Times by Integra Software Services Pvt. Ltd, Pondicherry, India.
Printed and bound in Great Britain by Antony Rowe Ltd, Chippenham, Wiltshire.
This book is printed on acid-free paper responsibly manufactured from sustainable forestry in which at least two
trees are planted for each one used for paper production.

Contents

8   Service Performance Optimization   **264**
Gerardo Gómez, Juan Torreblanca and Mattias Wahlqvist

List of Contributors

Mª Carmen Aguayo-Torres
Dpto. Ingeniería de Comunicaciones
Universidad de Málaga, Spain

Pablo Ameigeiras
Optimi Corporation, Spain

Brian Carroll
Aran Technologies, Ireland

John Cullen
O2, U.K.

Daniel Fernández
Optimi Corporation, Spain

Alejandro Gil
Optimi Corporation, Spain

Gerardo Gómez
Optimi Corporation, Spain

Juan Guerrero
Optimi Corporation, Spain

Petteri Hakalin
Optimi Corporation, Spain

Salvador Hierrezuelo
Optimi Corporation, Spain

Fergal Kelly (Foreword)
Vodafone, Spain

Manuel Martínez
Optimi Corporation, Spain

Héctor Montes
Optimi Corporation, Spain

Jorge Navarro
Optimi Corporation, Spain

Juan Ramiro-Moreno
Optimi Corporation, Spain

Gabriel Ramos
Optimi Corporation, Spain

Raquel Rodríguez
Optimi Corporation, Spain

Rafael Sánchez
Optimi Corporation, Spain

Pablo Tapia
Optimi Corporation, Spain

Juan Torreblanca
Optimi Corporation, Spain

Mattias Wahlqvist
Ericsson España S.A., Spain

Foreword

The ***Customer is King*** is a phrase I believe many people will be familiar with and one that I believe is highly relevant to the topic of Service Performance and Optimization. Brought back to its most fundamental driver, our networks exist because we have customers and it is these customers that one way or another fund the development of our networks and our salaries!

As we have progressed from first generation systems through to third generation systems, things have become more sophisticated, largely due to the more ready availability of processing power which in turn has led us to rely on more complex modulation schemes and greater detail into the transport protocols. A side effect of this complexity can be directly seen in the number of configurable parameters per cell, increasing with an approximate order of magnitude across each generation step. The consequence of this vast configurability is that service optimization has become significantly more difficult.

Across the chapters of the book, the reader will be introduced to the various technological standards and their corresponding architectures, including their in-built mechanisms for performance management. Amongst these the opportunity for *Quality of Service* (often known as *QoS*) differentiation is introduced. Whilst QoS introduces an autonomous management of quality, the key word in my view is missing from its label – *differentiation*, as *differentiation* is the fundamental leverage achieved from that approach.

Later on, the reader is introduced to the concept of end-to-end quality of service management from an end-user perspective. For me this is good and bad. It is good in the sense that the end-user (also known as the customer) is considered albeit in aggregate form not individually but bad in the sense that the component-by-component approach whilst having some merit is, I believe, flawed.

In essence, the concept of component-by-component end-user end-to-end quality of service management is that, the system is made up of multiple components and that in order to manage the end-user quality, each of these must be assessed and compounded together to derive an overall view on performance. Whilst this is theoretically correct, and certainly can be used as an approach, it is in my view an ageing approach. The reason I say ageing is that the number of components that exist in a cellular network these days is huge and growing daily, so the task of identifying them and all of the changes becomes almost impossible. Given this, it is in my experience all too easy to omit components from the calculation of performance.

A complementary approach to try to overcome the main fallibility in the end-to-end approach is now emerging. Leveraging again the improvements in processing power and recognising that the system is an ever-evolving entity, the approach being taken at the leading edge is now to 'sniff' the interfaces between the key subsystems and to then correlate the transactions with the chains of activity relating to a particular transaction of an individual customer.

This interface sniffing approach gives two valuable data sets. The first data set relates to the concept of end-to-end quality of service management. By aggregating data by service-type, it is possible to see how the service is performing overall (i.e. the end-to-end view is achieved), as well as where precisely it may be failing. The second data set may be compiled by aggregating the data by individual customer. This approach is so far the only one that gives a true customer (end-user) perspective.

Having the true end-user perspective is a powerful tool. It allows us to do a number of things.

- To talk to our customers about the quality they are individually experiencing, not a network average that we have measured. Thus we improve credibility and customer satisfaction.
- To establish priorities in issue management, perhaps choosing to address highest revenue customer issues first, maximising revenue.
- To substantially increase revenues by identifying customer provisioning problems, releasing suppressed revenue. This issue is generally not caught by the component-by-component approach and in my experience can easily add 5% to revenues, in spite of the best of checks and controls!

If ever there was a motivation to conduct service performance and optimization, surely these must rank amongst the highest!

Fergal Kelly
CTO Vodafone, Spain

Preface

Introduction

We are plunged into the era of mobile services. Wireless connectivity worldwide is becoming a real 'need' in advanced societies, not only for business, but also for entertainment, communities or even security purposes. In that sense, a huge variety of services is nowadays coexisting in a very complex and heterogeneous network infrastructure, which is additionally managed by different parties. The *end-to-end Quality of Service* (QoS) is intended to achieve a seamless integration of the above-mentioned data services over the networks while providing the best possible experience to the end customers.

The complexity of monitoring and optimization processes of data service performance is quite evident, considering not only the quick diversification of emerging services and quality requirements associated with them, but also the coexistence of a huge variety of access technologies and the wide coverage through which such technologies are offered.

The service performance optimization should not be just seen as a need for upgrading the network with additional resources (over-provisioning), but to analyse the end-to-end scenario, service by service, in order to ensure a predefined service quality while minimizing the costs, i.e. optimize the network usage at the same time that customer experience is enhanced. This is the only way to minimize the unit cost of a call or data session.

The goal of a mobile network operator is to offer the customer an assured end-to-end QoS, with a variety of service levels and predictable service response. Such a goal requires a set of intra-domain (radio access and core networks) and inter-domain agreements, which is consistent along the end-to-end 'chain'. For that purpose, it is very important for mobile network operators and service providers to have the capability to measure the service performance as experienced by the customer, guarantee their QoS expectations and succeed in the launch of new mobile services.

Do not expect to find here a magic parameter or formula that solves all the problems in this respect. This book is aimed at providing a proper methodology as well as the technical background required for assessing and optimizing the *End-to-End Service Performance* over a multi-radio technology scope. A proper understanding of the key factors influencing the end-user performance is essential to fulfil such a goal.

We tried to approach these objectives by combining both theoretical background and analysis with practical examples obtained from simulations and live network measurements, which could illustrate realistically what a user may expect from current cellular systems.

Who Can Benefit from This Book?

Note that a very high number of entities and factors are involved in a mobile communication. Let us imagine a customer using his new 3G terminal for retrieving a real-time streaming video through the Web. In this simple scenario, an important group of experts in different areas must contribute to optimize the customer experience (terminal manufacturers, content providers, mobile network operators, network element vendors, etc.).

This book is intended to cover many different aspects in this end-to-end approach focusing on the performance perspective. Therefore, we are sure that a wide group of readers could benefit from this book. First, mobile network operators (whatever the access technology they are exploiting) may find useful guidelines and hints on how to assess, monitor, analyse and optimize the data service performance. Secondly, this book is also directed to mobile network vendors and developers, as well as telecommunications companies working on performance solutions, mobile applications and/or consulting services in this area. Finally, this book will also serve universities and other institutions with technical background on telecom networks as the main reference on QoS and service performance over wireless.

Content of the Book

The book is divided into 8 chapters, which can be summarized as follows.

Chapter 1 introduces the quick evolution that wireless technologies and services have been experiencing over the last years. This chapter also describes the motivation and the need for applying QoS mechanisms in mobile networks as a way to satisfy end-user experience and optimize network performance.

Chapter 2, 'Cellular Wireless Technologies', is aimed at providing an overview of current and future radio technologies from a performance point of view. It describes the different evolution paths and data capabilities to support emerging data services.

Chapter 3, 'Data Services Architecture and Standardization', describes mobile services architecture and functionalities, as well as introduces the reader to a particular set of services (including protocols, signaling and QoS requirements).

Chapter 4, 'Quality of Service Mechanisms', is aimed at providing an overview of what QoS stands for, describing which kind of mechanisms are currently available in IP and cellular networks, introducing the need for QoS differentiation and providing a classification of data services according to their QoS requirements.

Chapter 5, 'End-to-End Service Performance Analysis', presents the concept of *end-to-end performance* as the way to measure the quality of a network from the user point of view. Analysis methodology based on the cumulative effect of the different layers and network elements through the transmission path is also analysed. The focus is placed on a technology-independent point of view that allows analysing the performance of the services based on generic parameters.

Chapter 6, 'Service Performance Verification and Benchmarking', provides a detailed view of the performance indicators and measurement methodology for assessing the quality of a wireless data network. A combination of network indicators and service parameters provides the best understanding of the system status. Service performance benchmarking results from live mobile networks are given and analysed in detail.

Chapter 7, 'Customer Experience Management', is intended to outline a new approach to service management known as *Customer Experience Management* (CEM) or *Customer Centric Service Management* (CCSM) – a conceptual explanation of what CEM is; the reasons why CCSM is replacing traditional service management; and the impact on revenues and customer satisfaction will be covered in this chapter.

Chapter 8, 'Service Performance Optimization', is intended to provide some guidelines on how to optimize the service performance, describing the main techniques available for that purpose and quantifying the performance gain for some of them.

Gerardo Gómez and Rafael Sánchez
The Editors

Acknowledgements

First of all, we would like to thank all of the people that directly or indirectly have contributed to the publication of this book. We are really proud of the whole team that has allowed this challenge to become a reality despite the great workload, full of tight schedules, trips and weekends of work.

Many thanks to Juan Melero for providing the opportunity to start this project and putting us in contact with John Wiley & Sons, Ltd, in addition to his great advice whenever we needed it. Thanks also to Mark Keenan for his guidance and help to contact relevant people in the industry whose input makes this book closer to operators and field engineers. It has been a real pleasure to include the contribution from Fergal Kelly as the Foreword.

This book would not have been possible without the contribution of all our good partners who gave their experience and support in many different areas: Pablo Tapia, Juan Ramiro-Moreno, Petteri Hakalin, Raquel Rodríguez, Juan Torreblanca, Salvador Hierrezuelo, Alejandro Gil, Juan Guerrero, Daniel Fernández, Pablo Ameigeiras, Jorge Navarro, Gabriel Ramos, John Cullen, Héctor Montes, Manuel Martínez, Mª Carmen Aguayo-Torres, Brian Carroll and Julia Martínez. With special care, we would like to express our recognition to Mattias Wahlqvist for his incredible help, support and planning capacity.

And of course, our immeasurable gratitude to all the colleagues who have supported, and even paid theirselves, our measurements in the two continents and many different cities: Juan Torreblanca, Juan Guerrero, Juan Pablo Iriarte, Miguel Ángel Álvarez, Timothy Paul, Héctor Montes, Greg Evans, Jay Langford and David Paolini. Special thanks to Salvador Hierrezuelo who not only contributed with many measurements and great analysis, but also took care of all the technical preparations and software fine-tuning, which made it possible to perform all the tests.

Not forgetting the great help from Optimi Corporation, allowing the usage of its software and always providing us with its full technical support and expertise in anything we needed, and all the operators and other companies we have worked with over the last years and which have provided us with the experience needed to face this challenge: Nokia, Cingular Wireless, AT&T, TIM, Telefonica, Vodafone, and many others.

Many thanks to the John Wiley & Sons, Ltd, publishing team (Mark Hammond, Sarah Hinton and Olivia Barnett) for their support during the development of this book.

This work is part of the research and development activities of Tartessos Technologies S.A. (Optimi Corporation). These activities are partially supported by 'Consejería de Empleo y Desarrollo Tecnológico of the Junta de Andalucía'.

Finally, we would like to thank our families and friends for their patience and love.

I, Gerardo Gómez, would like to thank my parents, my brother and the rest of my family, and especially express my loving thanks to Miriam.

I, Rafael Sánchez, would like to thank my parents and brothers who have always been there. Thanks to my close friends who always push for a break – those who are far away and never forget – and to the little baby-girl, Lucía.

We welcome any comment or question you may have related to this book in order to continue improving it. The e-mail address used for this purpose is: qos_book@hotmail.com.

Gerardo Gómez and Rafael Sánchez
The Editors

1

Introduction

John Cullen, Mattias Wahlqvist and Gerardo Gómez

1.1 Mobile Services in Perspective

Twenty years ago mobile phones were a rarity with less than 5 million subscribers world-wide. They tended to be fitted to cars as car phones as they were bulky and power hungry, used by the elite due to the high prices charged for equipment and service, provided only voice call capabilities and only delivered service over what we would consider a small area today. At the same time, even those companies launching mobile services predicted that the overall market would be very small. Ten years later, many industry observers still believed that the market would remain relatively small.

Today, mobile devices are used by around 1.5 billion people worldwide, a three-hundred-fold increase since 1985, which equates to a worldwide penetration slightly over 20%. Mobile communications is now a technology for everyone. For many people it is now an indispensable part of their life with their mobile being among their key personal possessions alongside their watch and wallet.

The mobile device has changed all our lives and the way we live it. Listed below are a number of examples.

- *Mobility*: Today, we are travelling more for both business and leisure. This has lead to a heavier reliance on the mobile phone to stay in touch with colleagues and friends/family.
- *Planning*: As everyone is reachable, we do not plan ahead. How many times have you heard: 'Yes. Let's meet at 12 in the city centre. I'll call you when I arrive, so that we find each other.'
- *Communities*: Part of the tremendous popularity of mobile devices is that wherever you are, communication-wise you are very close to your friends and colleagues. Teenagers today rely heavily on their mobiles to keep in touch with their friends and to organize their social lives. To do this they heavily rely on text to communicate with their community.

End-to-End Quality of Service over Cellular Networks: Data Services Performance and Optimization in 2G/3G
Edited by G. Gómez and R. Sánchez   © 2005 John Wiley & Sons, Ltd

- *Participation TV*: TV shows are trying to appeal more to their audiences by allowing their audiences to interact with their shows so as to affect the outcome of the show (e.g. reality TV shows) or generate content (dating/chat shows) while also providing a revenue generation opportunity.
- *Marketing*: Many consumer brands have started launching competitions or offers whereby entries are made via SMS and an instant response can be given to customers. In some cases, prizes are downloads for handsets that allow customers to personalize their handsets with ring tones or wallpapers. At the same time, the consumer brands are able to build up marketing databases using entry information.
- *Security*: Today, most of us would not think of travelling long distances in a car without having a mobile with us in case of an emergency. Also today, in richer countries, many parents are giving their young children mobile phones so that their children can contact them in an emergency and so that they can keep track of their children.

For many young people today, their first commercial relationship with any communications company is with their mobile operator. For this wireless generation, the mobile is at the hub of their social lives. When they first move away from home, they maintain a relationship with their mobile and in most cases the mobile service becomes the only communications service they subscribe to themselves. As a consequence, their relationship with a mobile operator is their prime relationship with the communications industry replacing the traditional prime relationship enjoyed by fixed operators. Today, this unique relationship with the mobile industry tends to be broken only when an individual moves into their own property and starts to consume services that require fixed lines such as broadband Internet services.

Looking forward, we are setting out as an industry on a new phase of market development where with Third Generation (3G) radio technologies the number of services and the richness of those services is greatly expanded. Five years ago, the mobile industry talked about the highest data rates that would be available for 3G. These high data rates are still an issue for laptop PC users with data cards, but for average handset users 3G opens up the ability to use new richer services and capacity that would not have been possible for mass-market customers with Second Generation (2G) technologies. Listed below are a number of examples of how mobiles could be changing our lives in the future.

- *Communication*: Video calling is starting to allow consumers to communicate face to face and to share their environment with their colleagues. In today's busy world with frequent travel, it allows families to keep in touch while on the move.
- *Entertainment*: Music download and streaming is allowing people to get and listen to music on the move, releasing them from computers and fixed communications. At the same time, the ability to download games, which is possible today, will be enhanced by the capability to play them on the move with online friends so providing a new dimension to gaming.
- *Current affairs*: Already the first 3G operators are offering consumers the ability to keep up with events on the move via video clips so allowing consumers to be able to see, for example, their team winning a sports game while on the move.
- *Content creation*: The emergence of smart phones incorporating cameras, good quality displays and reasonable processing capabilities will allow consumers to create and share content. Content could be owner-generated pictures, videos, audio, text or any combination

of these media types. Sharing could be by picture/video messaging, via online electronic journals (blogs) or by peer-to-peer file sharing. To safeguard personal content, network backup capabilities will become essential.

- *Purchasing*: The arrival of large colour displays on devices will make it more practical for consumers to buy services from the Internet and carry out transactions on the move so freeing consumers from their fixed PCs and allowing them to make use of dead time when travelling, waiting for friends, etc. It will also provide a rich channel for governments to communicate and interact with their citizens.
- *Business*: On our company networks today we have from our PCs high-speed access to company resources and to the Internet. On the move, our PC connectivity has been limited by either connection speeds or the availability of hotspot coverage. The emergence of 3G technologies will enable us to improve this situation by providing coverage over large geographic areas.

Like the Internet world, the success of mobile data services will be built for giving consumers access to a rich set of services so as to satisfy a multitude of customer needs at the right cost. Unlike the Internet world, in the mobile environment the winning services and service providers will be determined not only by the simplicity of using services on the move but also by the quality of the experience in using services – the best service in the world will not sell if a user needs an answer in ten seconds and it takes one hour. This book aims to look at how the service performance can be tailored to give the right performance at the right price.

1.2 Mobile Technology Evolution

Today, mobile telephony is a global industry with a global footprint in a large part of the populated world. In the beginning, however, mobile telephony systems were typically a local solution on country level.

1.2.1 Reasons for Mobile Technology Evolution

There has been a tremendous evolvement of mobile telephony during the last 20 years, both technology-wise and service-wise. One interesting aspect of the evolution of mobile technologies is to ask yourself what is really the driving force being the engine for the switchover from one technology to another. That is a complex question, and there is not one true answer. It is also so that the answer will depend on whom you ask. Here we anyhow try to illustrate the complexity of this question by giving a few opinions from different points of view.

- *Customer service requirements*: Is it so that end-users are demanding better and more requiring services, which leads operators and vendors to implement new technologies? This statement is partly true and it is important to observe that it will likely become truer as time passes. In the beginning, the mobile telephone service was just a telephone service you could use on the go. Today, there are additional services (SMS, WAP etc.) that are adding new requirements to the system. It is also so that the end-users today are much more advanced in terms of comparisons with, for example, services on the fixed Internet.

If a person can download a large email on the fixed computer, there should be no reason why he/she should not be able to do it in his/her mobile phone.

- *Customer and traffic growth*: Is it so that the growth in the customer base and the traffic that generates are implying that the operators need to reinvest in newer more efficient systems? This is not really true. Typically, new features (e.g. half-rate codecs, frequency hopping etc.) are introduced to enhance capacity and quality, but it is of course important for the operator to protect his CAPEX investment as long as possible. It is also so that the time to design a new system makes it impossible to rely on a new more spectrum efficient handling of the traffic. The problems are here today, and the future system will take many years to get into the field.

- *Differentiation of services and Quality of Service (QoS)*: Is it so that new systems are developed to be able to perform service differentiation and offer QoS? To some extent yes. It is a common understanding that service differentiation and QoS is the only way to cost-efficiently offer a wide range of services. Still, the service differentiation has already been gradually introduced in today's systems, and so making service differentiation a main reason for the development of new systems is only partly true.

- *Spectrum availability*: When new spectrum is made available there is of course an urge to make use of it in the best possible way. Spectrum is a scarce natural resource, and the introduction of new more efficient systems is done easily if it is introduced together with a new spectrum band.

- *End-user requirements*: The end customer has normally a firm opinion on whether he likes a service or not ('like' in this context normally means that he thinks that it is worth paying the stipulated price for getting the service). That opinion heavily affects his usage of the service. Still, considering the time it takes for a service to become a mass-market service, makes us believe that it is not end-user requirements that are driving the need for new systems. The majority of the end-users are not advanced enough to know what they will need in a five-year time frame.

- *Commercial aspects*: There are of course commercial aspects that influence the willingness to introduce new systems into the markets. Vendors might want to protect or increase their market share; operators might want to create a high-end profile towards their end customers etc. Considering the time frame to introduce new systems, it is anyhow clear that the commercial aspects are mainly considered on high strategic level.

To conclude, we can see that there is a variety of reasons for new mobile systems to be introduced, with the strongest ones being the need to make more and more efficient use of a limited natural resource, the spectrum. On top of that, there are a multitude of other reasons to consider.

1.2.2 Mobile Technology Evolution Paths

Analog technologies were dominant in the cellular market up to 1997, when their global market share was exceeded by that of 2G digital technologies. From that date, the Global System for Mobile communication (GSM) revolutionarily changed the way we look at and think of mobile telephony. After its introduction we have seen a rapid evolution of services, technologies and performance. GSM technology's market share shows a sustained growth and today it has become the global 2G standard, deployed by more than 460 operators around the world and accounting for more than 70% of the total number of cellular subscribers.

General Packet Radio Service (GPRS) technology, developed as a Packet Switched (PS) extension of the GSM network, allowed high-speed access to IP-based services and at the same time it provided an efficient use of the network resources. Some time later, Enhanced Data for Global Evolution (EDGE) technology increased the radio data rates by including some enhancements in the modulation and coding schemes. (E)GPRS can be considered as the convergence point between the Time Division Multiple Access (TDMA) developed in North America and GSM technologies, and is the foundation for the PS domain of the 3G Universal Mobile Telecommunications System (UMTS).

Another parallel technology evolution path is the one coming from cdmaOne. Despite an important growth during its first year of deployment, cdmaOne's (and its main successors: CDMA2000-based family) market share has stabilized around 15% of market share in 2004. Although a natural evolution from CDMA2000–1x would be the support of 1xEV-DO (1xEvolution, Data Optimized) and 1xEV-DV (1xEvolution, Data and Voice), many CDMA operators are currently migrating towards GPRS and EDGE technologies as an alternative option (with the later integration of WCDMA). This last option is however dependent on the cellular-operator's licensed bandwidth, since WCDMA technology is currently not supported in the 800-MHz band, the future availability of dual mode cdmaOne/WCDMA terminals and the integration effectiveness of the technologies.

Expected WCDMA launch is nowadays becoming a reality as the evolution path of 2G technologies, being already supported over several markets around the world. The convergence of 2G technologies towards the UMTS multi-radio 3G evolution path is clear. The entities, such as operators, global associations and standardization bodies, which are representing and driving the evolution of three out of the four current most representative 2G technologies, have endorsed the UMTS multi-radio evolution path. Figure 1.1 summarizes the evolution paths associated with the existing 2G technologies.

The evolution of the mobile technology's market share and how these technologies are distributed around the world is depicted in Figures 1.2 and 1.3, respectively [1].

Thanks to the evolution of the networks towards PS technologies, data services have experienced a huge increase in terms of data transmission capabilities, leading to an important increment in operator revenues. Currently, SMS and MMS are still the most profitable,

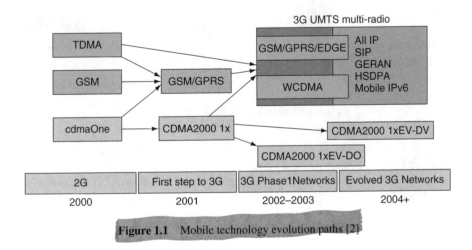

Figure 1.1   Mobile technology evolution paths [2]

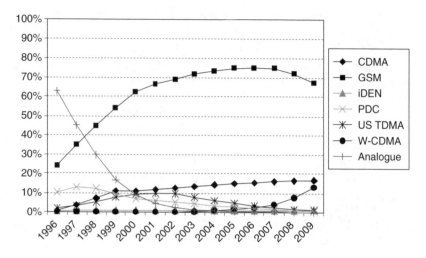

Figure 1.2   Mobile technology's market share (forecasted from 2005 onwards) [1]

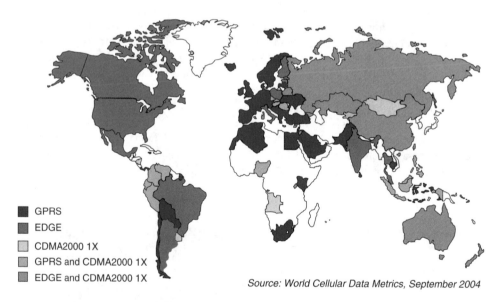

Figure 1.3   Mobile technology deployment [1]

although other services like email, content downloading (i.e. Java applications, games, tones, etc.) or streaming are already pushing hard.

SMS and MMS, together with ring tones and information downloads, have represented between 2 and 7% of operator's revenue in both North American and Latin American regions in Q2 2004. China Mobile handled more SMS than any other operator, 30.9 billion in Q2 2004 [1]. Although in those regions, CDMA2000–1x was the most widely deployed technology, data usage has been boosted by the continued deployment of advanced data networks (GPRS/EDGE), as shown in Figure 1.3.

In Europe, GPRS and partially WCDMA have been deployed until today, where an average data percentage of revenue reached 13.6% in Q2 2004. SMS traffic in western Europe grew approximately 17–18% in the 12 months by the end of June 2004 [1].

MMS had been launched commercially by 237 operators in 88 countries in September 2004. MMS usage and traffic volumes on the whole remain low, being KTF Korea and Verizon (USA) the ones reporting a higher number of MMS (over 21 million in Q2 2004).

Total mobile subscribers to GPRS, CDMA2000-1x, I-mode and other advanced data services exceeded the 150 million mark in Q2 2004, and the total reached just over 152 million as at 30 June 2004, or 9.9% of the world's total mobile users. The reader is kindly referred to Chapter 2 for a detailed description of the different technologies listed along this section.

1.2.3 Harmonization/Evolution Challenges

With the design and commercialization of a new system, there are a large amount of requirements that need to be considered not only from a technical perspective, but also from an economical and commercial point of view. Here we list a few major challenges that are important to consider.

- *Backwards compatibility*: In order to get maximum reuse of older investments, a natural evolution also requires the new system to be backwards compatible towards older, already commercially deployed systems. For example, with the introduction of UMTS in Europe, it is of high importance that already from day 1 being able to perform inter-system handovers to and from the commercial deployed GSM network. The reason for this is obvious, as the operator wants to be able to offer continuous service coverage. Note, however, that general requirements on backwards compatibility can mean many very different technical requirements on the new system. An operator might want to make maximum use of his already deployed network, which might lead to, for example, requirements on inter-system handovers and the ability to co-site the two systems. On the other hand, a mobile phone vendor might want to ease the implementation of a multi-system handset, and might want to set requirements on clock frequencies to be used, as well as limiting the complexity between how the two systems interact.

- *Service transparency*: New systems typically offer new services. For that reason it can be difficult, especially in the beginning, for the operator to offer a continuous service support over the whole coverage area. For example, high-speed data service is impossible to maintain in UMTS when the user leaves the UMTS coverage area and are handed over to GPRS. From this aspect, it is anyhow considered important to be able to maintain some type of service, even if the service level is lower. Whether that is useful or not for the end-user is anyhow very service dependent.

- *Interoperability/roaming*: As users move between systems, it is difficult for the operator to maintain a constant service level. In the same case as in the service transparency example above, a user that is roaming into another network might not get all the services he/she can normally access in his/her home network, as they might simply not be implemented. Another consequence is the implications this might have on the billing models to be used.

1.2.4 Future Outlook

It is clear that the development will not stop here and now. New systems and features are today being standardized and developed for inclusion in the upcoming years. A global traffic growth together with the release of new spectrum or reforming of old spectrum will also increase the need for newer more efficient ways to transport mobile data communication. Lately, a few trends have however emerged that might have significant impact on how future mobile communication systems are designed and deployed, although they are not going to change the evolution path.

- *Emerging markets*: There are still large parts of the world where mobile communication has not yet been deployed and there is a large market potential. Typical situation for these countries is that they have a high potential subscriber base, and that the fixed phone infrastructure is not so developed. What holds back a massive deployment in these countries is normally that the average amount that can be charged to the subscribers is relatively low, which makes it challenging to deploy and market a network in a cost-productive way. Ultimately, this might lead to requirements to develop 'low-end' systems with lower production cost and less features.
- *New services*: With the exponential growth of the Internet, there is also an explosive amount of new services that users can access. By getting used to access these services on the fixed Internet, there will be a demand for doing at least some of them while being mobile. In addition, mobile communication can offer a multitude of mobile services that will also add new requirements on any future system.
- *New users*: The introduction of mobile data services also opens up for a complete new subscriber group – machines. The vending machine can itself send its order for new drinks or to get service, or you could remotely find your car on a map and demobilize it when it gets stolen. This is an area becoming more and more important with an infinite amount of possibilities that we likely will see.

1.3 Motivation for QoS

The motivation to look at QoS is two-fold.

1. To provide a service experience to consumers that meets their expectations so that they are more likely to use it again and recommend it to friends or colleagues.
2. To achieve optimum loading of an operator's network so that the desired service experience is delivered for each customer while maximizing network utilization.

The following section provides a brief introduction to the main factors involved in addressing these two issues.

1.3.1 Service Experience

In the early data services market, many consumers are impressed to just use a service when mobile. However, once this euphoria has passed, the vast majority of consumers start to judge a service based on how it meets their needs and expectations. As an example, the Short

Message Service (SMS) was designed as a store and forward service, and was offered as such by most operators in the mid-1990s with messages sometimes delayed hours before delivery. Consumers have conclusively taken to the service as a way of communicating quickly and efficiently with friends and colleagues without the need for a conversation. However, in meeting these needs we expect that a message is delivered almost instantly. When important business and social meetings are arranged by SMS, a couple of hours delivery delay is unacceptable. As we can see from this example, a consumer's expectations of a service dictate whether it is perceived as working well or badly. These expectations in turn determine the critical success factors that the network must deliver against if the service is to be perceived as good. Take the following examples.

1. A customer using an 'always on' email application (e.g. Blackberry) expects their emails to be accurately received and that they are received within a reasonable time, for example 10–20 minutes after being sent. This implies that the network must deliver accurate information, i.e. a very low bit error rate, but that the payload can be delayed for a reasonable amount of time.
2. A customer using a Push-to-Talk application will expect to get voice messages within a couple of seconds from their friend or colleague sending a voice message but must be prepared to tolerate some voice distortion on limited occasions. This implies that the network must expedite the voice messages through the network but that limited packet loss can be tolerated.
3. A customer browsing the Internet from a laptop PC with a 3G data card will expect that the Web page loads accurately to a point where they can start reading in less than 10 seconds, otherwise their concentration will lapse making the service uncomfortable to use. This implies that the network must deliver accurate information and that some limited delay can be tolerated. In this example, the way the Web page is built can also make a difference. For example, a Website that displays text within 10 seconds but then downloads images in the next 10 seconds will often appear to be quicker than a page that completely downloads in only 15 seconds.

In all these examples, if the network achieves the critical success factors, the consumer is likely to perceive a service as working well. If the network fails to deliver, the consumer is likely to perceive a service as working poorly.

When considering the quality of a network it is worth remembering that services run either between two terminals or a terminal and a server, and that the critical success factors apply across the whole connection. There is no point, for example, in engineering the GPRS network to meet the consumer's expectations when the connection to the content provider is not to the same standard and hence degrades the overall experience.

The main causes of a network failing to deliver against the critical success factors are:

- *Radio network performance* – Are there a lot of errors on the radio interface?
- *Network capacity* – Is there sufficient capacity to deliver a good service?
- *Network design* – Is there too much delay in the system; is sufficient capacity available end to end?
- *Application design* – Are the right protocols being used for a mobile environment?
- *Service support* – Is service enhancement technology correctly configured?

1.3.2 Radio Network Performance

A well-planned radio network where data errors on the air interface are minimized in most cases will improve application performance. If there are a large amount of errors, retransmissions are required which can slow down the amount of information that can be transferred by protocols such as TCP. Radio errors also introduce extra delay into any conversation between application clients/servers, slowing down application response times. In a similar way, voice conversations can be slowed down by satellite delays.

1.3.3 Network Capacity

Well-designed mobile networks are dimensioned so that they have just enough capacity during busy periods. Any more capacity than necessary adds network cost for operators reducing profit margins. Too little capacity and customer applications will not be able to get network capacity so will deliver a poor customer experience. It is, however, extremely difficult to predict loading accurately as demand may fluctuate by time, day, month and season as well as demand growing with time. As a result, any operator wishing to offer customers a good experience, so as to encourage the uptake of services, would have to expensively over dimension their network to avoid congestion. To overcome this problem, standards have defined the concept of QoS and this is starting to be implemented into network equipment. The QoS concept encapsulates the idea that different data streams could be treated differently by the network depending upon the service being carried (Chapter 4). Ideally, a service that requires fast response time is assigned a QoS that in periods of congestion it would receive priority over other traffic. Conversely, a service that can tolerate a reasonable delay would have lower priority than other traffic. By assigning different QoS to different services, when congestion occurs traffic can effectively be smoothed over time with high priority traffic still being transmitted with little delay but lower priority traffic being delayed until capacity is available. As a result, in periods of temporary congestion, a network providing QoS can meet customer-service expectations with existing capacity. In this way, with QoS, operators can more effectively load their networks as they can tolerate temporary congestion while at the same time ensuring that they deliver customers with the good service experience they expect.

It is worth noting, however, that QoS mechanisms are designed to work in periods of temporary congestion, where lower priority traffic can be delayed without impacting the service experience of those services. If heavy congestion occurs or the network is congested for extended periods of time, QoS cannot be relied on to maintain the customer-service experience. In these cases, further capacity is required and the network should be re-dimensioned to a level where only temporary congestion occurs.

1.3.4 Network Design

The principle aspects of network design that can impact service performance are:

- *System delays* – Every additional piece of end-to-end delay slows down application 'conversations'. For services where customers demand low response times, these delays can make a service unusable if the application requires an extensive 'conversation'.

- *System design* – An operator may choose to set up a data session when the phone is switched on, requiring costly resources to be allocated as long as the phone is on, or when a service is activated, adding additional delay while the data session is set up.
- *Equipment* – One node will always be the system bottleneck. It is important to understand where bottlenecks may develop and understand the scenarios in which they will occur. For example, a router could be limited at any point in time by the number of active sessions, the data throughput per second or the number of packets it can transfer in a second. In this example, changing the mix of services in the network from predominant Web browsing to Voice over IP could change the place and type of bottleneck.

More information on the impact of service performance on design can be found in Chapter 5.

1.3.5 Application Design

Application design can impact on service performance in two ways. First, the user-interface can be designed so as to soften the impact of network performance on the user, as described in the Web page example earlier. Secondly, the application can be designed so as to better carry out communications over wireless connections. This book leaves the issue of user-interface design to usability experts but instead chooses to focus on application design. In focusing on application design, particular attention should be paid to:

- *Protocols used* – TCP assumes that packets that do not arrive in a certain time period are lost due to congestion and therefore the amount of information that can be transmitted at any one time should be limited. In radio networks, these 'lost packets' may have nothing to do with congestion but could be the result of a temporary radio failure lasting milliseconds. As a result, protocols used to build new services should be considered carefully.
- *'Conversation' structure* – In order to deliver a service, application clients and servers tend to have a dialogue so as to transfer information on the service requested, the application/person requesting, what format the service should be delivered in, etc. If the dialogue is sequential, one bit of information is transferred and acknowledged before the next bit can be transferred and acknowledged. In mobile networks with high delays the overall effect of the delays in multiplicative. This leads to unacceptable service delays from a customer perspective. Ideally, for good performance as many actions are performed in parallel as possible. One consequence of this is that when using HTTP, which provides transport for Web page requests and responses, version 1.1 considerably outperforms, by up to a factor of 3 times, version 1.0 simply because it allows parallel processing of Web page components.

More information about impact and interactions between the wireless system and different applications can be found in Chapters 3 and 5.

1.3.6 Service-Enhancing Technology

The aim of service-enhancing technologies is to improve the performance of data services by overcoming limitations in wireless systems caused by the radio environment. Examples of service-enhancing technology are:

- *Payload compression* – Information transferred across the radio interface is compressed first so that it takes less time to transfer over a radio connection.
- *Controlled quality degradation* – Higher quality and resolution especially of pictures implies larger file sizes and thus longer transmission times. When the final end of the transmission is a mobile terminal with limited screen resolution, in addition to normal compression, images can be downgraded or reduced in size so that they are still recognizable and fit better to terminal screens. This process is a compression with losses and can perform in network elements such as proxies.
- *Proxies* – Proxies store copies of Web page components locally so that when a Web page is requested, the proxy can instantly provide the Web page components without incurring further delay as it avoids having to get the Web page from the Internet. As Web pages can be made up of many components which need serial requesting, cutting the end-to-end delay associated with collecting components can have a significant impact on Web page download times.
- *Protocol optimization* – Protocols are adapted to make them more suitable for use in the radio environment. The most common optimization is related to the TCP protocol which is used to carry the majority of Web traffic. The optimization aims to counteract the reduction in flow rate associated with delays to TCP acknowledgements, which TCP interprets as congestion, by injecting fake acknowledgements into data streams. At the application level, optimization can take the form of consolidating all the Web components into one transaction, which is downloaded so that long delays caused by requesting and receiving Web page components serially is avoided.

Service-enhancing technologies are usually implemented together with performance enhancing proxies, which are treated with more details in Chapter 8.

1.3.7 Conclusion

The aim of service performance optimization is to provide a service experience for customers that meet their expectations while maximizing network utilization. Providing a good service performance requires operators to examine and optimize how they deliver services at multiple network and system levels. Today, many operators spend significant effort in optimizing their voice networks. Optimizing service performance for data services is likely to be a significantly more complex and resource-consuming task as there are considerably more variables. The following chapters of this book will examine each of the issues raised in this section in considerably more detail so as to provide guidance on how to achieve the best service performance in mobile networks.

References

[1] EMC Database: www.emc-database.com.
[2] T. Halonen, J. Romero and J. Melero, 'GSM, GPRS and EDGE performance', Ed. Wiley, 2nd edition, 2003.

2

Cellular Wireless Technologies

Petteri Hakalin, Pablo Tapia, Juan Ramiro-Moreno, Raquel Rodríguez,
Mª Carmen Aguayo-Torres and Rafael Sánchez

2.1 Introduction

The wireless and cellular scenario is still nowadays a mixture of different technologies which coexist and compete in the same markets providing a variety of different services, in what is called Second Generation (2G) of mobile communications. However, thanks to the work of the different standardization organizations, the tendency is to evolve and unify these different systems into one common system, which would allow full seamless communication and mobility to the users.

Still for the new Third Generation (3G) technologies, aimed to provide wideband and high-speed services for voice and data through cellular networks, there are different evolution paths: the GSM/GPRS/UMTS path, promoted by the Third Generation Partnership Project (3GPP), and the CDMA/1x/EVDO path, which is promoted by 3GPP2. Both systems are competing for providing 3G capabilities and services for the end-users.

Despite discussion on which solution may provide the best results, the objective of this section is to provide a general overview about the systems involved in these two evolution paths, including a brief description of the architecture and capabilities. In addition to 2G and 3G cellular systems, other possibilities currently deployed for wireless solutions, such as WLAN will also be analyzed, considering the main benefits and peculiarities that these technologies present.

Finally, an introduction to the future Fourth Generation (4G) systems, seen as a full integration of all the coexisting systems that can be found nowadays will be presented. It will be intended to show the main lines of investigation to improve the performance and provide high bandwidth for demanding services through mobile terminals.

As a starting point, Figure 2.1 presents a classification of different systems currently available, according to its mobility scheme and available bandwidth. The classification not only shows cellular systems, but also other technologies, wireless and wire-line, which compete in performance, speed and availability with both 3G evolution path.

End-to-End Quality of Service over Cellular Networks: Data Services Performance and Optimization in 2G/3G
Edited by G. Gómez and R. Sánchez © 2005 John Wiley & Sons, Ltd

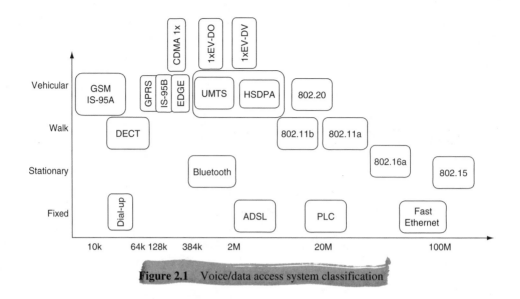

Figure 2.1   Voice/data access system classification

2.2 GSM/GPRS/EDGE

The GSM standard (Global System Mobile) was conceived in the 1980s by the European Telecommunication Standards Institute (ETSI) as a digital mobile system that would substitute the existing analog telephony (NMT, AMPS, etc.). The spirit of the standard was to create a system that could carry higher capacity, trying to ensure a unification of the mobile telephony in the European zone and therefore facilitating subscriber movement into the different countries (roaming). At the same time, other similar digital mobile alternatives were created along the world, such as IS-36 in America or PHS in Japan.

The GSM standard has been a success in the so-called 2G technologies, being adopted by many different non-European countries around the world and even substituting in some cases existing 2G networks, as it is the case of some major operators in the US and Latin America. Nowadays, GSM accounts approximately for 80% of the world's 2G networks. The open protocols strategy, as well as a continued improvement of the system with enhancement features such as Frequency Hopping, Adaptive Multi-Rate (AMR), and in the current times Single Antenna Interference Cancellation (SAIC) have been very important for this success. The original GSM has evolved into a family of standards (3GPP) that includes several technologies, which have been designed to be coexisting as complementary systems. These are the cases of GPRS, EDGE, WCDMA (FDD and TDD) and TD-SCDMA.

In 1997, following the increasing use of the Internet and the explosion of the data communications, the ETSI developed the General Packet Radio System (GPRS), a mobile packet data system which could easily be deployed in GSM networks with minor impacts on the architecture and network elements. This system was conceived to carry small amounts of data at relatively low bit rates (typically from 30 to 40 kbps) and being used as transition to high-speed mobile data networks (3G), which at that time were starting to be designed.

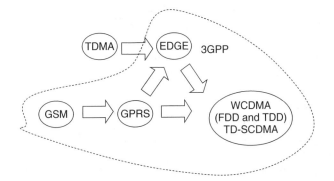

Figure 2.2   3GPP path evolution in the Americas

Later on, improvements in the GSM radio interface, together with minor changes to the GPRS protocols created EGPRS,[1] extension of the system to an intermediate level between 2G and 3G capabilities. Depending on the terminal capabilities, EGPRS is able to provide data rates from 50 to 220 kbps, with legacy devices (up to 4 TSLs (timeslots)), and potentially up to 440 kbps with 8 TSLs terminals. The introduction of the EDGE technology facilitated the possibility to merge American TDMA networks (TIA/EIA) into the GSM evolution path, with the intention to gain the American markets for the 3GPP 3G solution (Figure 2.2).

In the 1990s, different standardization institutions from around the world decided to join their efforts in the design of the next generation mobile telephony, the UMTS project. This association was named 3GPP and it was composed of several standardization groups: TTC & ARIB from Japan, ETSI from Europe, T1 from USA, TTA from Korea and CWTS from China. The GSM/GPRS system was selected as the main path for the migration to 3G and it was therefore included in the 3GPP project, under the name GERAN (GPRS/EDGE Radio Access Network). The core network part is basically the same in 2G and 3G networks.

There are several releases for the 3GPP GERAN system: R4, R5 and lately R6. Most of the equipment available in current networks will be compliant to R99 or even older ones, still specified at ETSI group.

2.2.1 Description of the GSM System

2.2.1.1 Deployment

The GSM system is a digital mobile telephony system that uses multicarrier Frequency Division Duplex (FDD) and Time Division Multiple Access (TDMA) configuration, and can operate in the different frequency bands: 400 MHz (Europe only), 900 and 1800 MHz (typical rollout), and 850 and 1900 MHz (used in the Americas due to spectrum constraints).

---

[1] EGPRS is also widely known as EDGE. EDGE (Enhanced Data Rates for GSM Evolution), is the acronym used to refer to a series of physical layer improvements introduced in the GSM system.

The information sent to and from the mobile station is transmitted into separate frequency channels, which have a constant separation depending on the band in use.

The different carriers are spaced 200 kHz, whereas the TDMA frame is divided into 8 slots, each of them with a length of 0.577 ms. In the original GSM system, the modulation used was GMSK, which allows the transmission of 1 bit/symbol, and provides raw data rates of up to 9.6 kbps for voice channels. In the case of data channels, as it will be seen later in this chapter, the raw data rates can raise up to 22.8 kbps per TSL.

Due to the multicarrier nature of the GSM system, the frequencies available have to be carefully distributed in order to avoid zones with high interference. This is achieved by means of frequency plans which are characterized by a certain average reuse distance. Several standard patterns exist for frequency plan strategies based on regular grid layouts: 1/3 (reuses the same frequency every third cell, as depicted in Figure 2.3), 2/6, 3/9, 4/12, 5/15, etc. However in most of the cases *ad hoc* frequency plans have to be implemented taking into account the specific particularities of the zone where the network is located.

Nowadays the distribution of frequencies in GSM systems has changed since the introduction of Frequency Hopping (FH) techniques. This mechanism improves the performance of the fixed frequency plans due to the gains associated to frequency and interference diversity and allows tighter frequency reuses, which in practice means higher system capacity.

Other capacity enhancement techniques adopted by GSM systems are Power Control (PC) and Discontinuous Transmission (DTX). The first one aims at reducing the transmitted power per link in uplink (UL) and downlink (DL) directions based on a 'good enough' criteria and thus achieving a considerable interference reduction in the system. DTX feature avoids data transmission on the air interface during voice silent periods. The effect of DTX is double: on one hand, it is an effective way of saving power, which is especially important for the terminal battery; and on the other hand, the total amount of interference in the network is reduced to approximately a half, which enables higher network capacity.

With all these features active in the system (FH, DTX and PC), the GSM voice service can be deployed with relatively tight frequency reuses, achieving an effective reuse of 6 and even lower. However, these techniques cannot be applied to all the channels in the GSM system, which causes an overall capacity reduction, especially in networks with small allocations of spectrum.

Different logical channels are defined in the GSM standard, which are subdivided into traffic and control channels. Control channels can be dedicated to a user or be shared by multiple

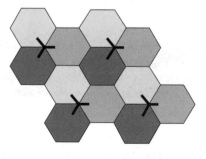

Figure 2.3   Ideal regular hexagonal grid for frequency planning

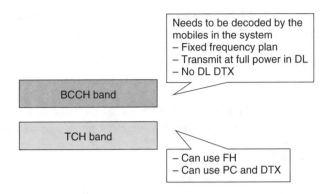

Figure 2.4   BCCH vs TCH layers in GSM system

users. Among the control channels, the Broadcast Channel (BCCH), the Frequency Correction Channel (FCH) and the Synchronization Channel (SCH) have special relevance because their configurations have a direct impact on the system capacity.

These special control channels have to be configured in a manner that they can be decoded by any mobile in the network that is near the cell, thus introducing the following constraints: (a) they need to be transmitted at maximum power, (b) they have to be deployed in carriers that do not use FH and (c) the reuse of the frequency plan has to be loose enough to enable a proper decoding of the broadcasted inform42ation, which in practice means using a reuse 3/9 or higher. These constraints will represent a reduction in the system capacity due to the fact that a part of the available spectrum needs to be deployed with a non-spectral efficient configuration (Figure 2.4).

2.2.1.2 Network Architecture

The GSM system is divided into three main units: the Base Station Subsystem (BSS), the Network and Switching System (NSS) and the Network Management System (NMS). While the NMS is a support system, whose scope is to help in the configuration and troubleshooting of the network, the BSS and the NSS deal with the transmission and switching of the connections, and are particularly important when analyzing the performance of the different services.

The main units to highlight in the basic GSM system (circuit-switched services) are shown in Figure 2.5.

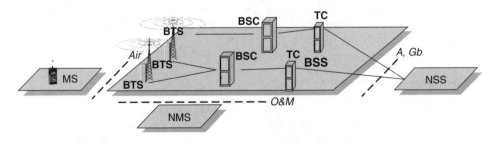

Figure 2.5   Basic GSM system architecture

- *Base Station Transceiver (BTS)*: These are the elements that define physically a cell, which can contain several Transceivers (TRXs). Each of the TRX transmits into 8 TSLs, which can hold one or two voice calls.
- *Base Station Controller (BSC)*: It is a switching and control center for a group of BTSs. Most of the Radio Resource Management (RRM) functionalities are implemented in this unit.
- *Mobile Services Switching Center (MSC)*: It has switching capabilities between different BSCs, as well as connection routes to other telecommunication networks (PLMN), and the network registers (VLR/HLR).
- *Visitor/Home Location Register (VLR/HLR)*: Contains information about the users belonging to the network and those that are roaming.

2.2.2 The GSM Transition to Packet-Switched Systems (GPRS)

In the 1990s, together with the idea of the 3G network, it was planned to evolve the existing GSM network, which was conceived of circuit-switched services, into a mixed voice and data system that could share some of the network elements that were to be deployed for 3G purposes.

With this objective, and trying to reuse as much as possible the existing infrastructure, it was conceived the General Packet Radio System (GPRS) for GSM. The changes introduced affected at different levels and network elements, but none of them modified the air interface.

2.2.2.1 New Coding Schemes for Packet Data

The packet-switched information is transmitted inside newly defined logical channels, named PDTCH (Packet Data Traffic Channels), which are encoded in a different manner than the voice traffic channels, but using the same TDMA and FDD structure. Four levels of protections are defined in a new codec set called Coding Schemes 1 to 4 (CS1 to CS4). The system is able to adapt the codec automatically based on radio conditions, being CS1 the most robust codec and therefore the one that can be used under worse conditions and CS4 the least protected codec, which can transmit at a higher bit rate if radio conditions are good enough. With this codification, the GPRS connections can achieve up to 20 kbps per TSL.

Another important change introduced with GPRS is the multislot configuration, where the data users can transmit information simultaneously using several TSLs from the same carrier, and multiplexing, which allows that several users dynamically share the same TSLs. The standard defines limitations to the number of slots that can be used at the same time by a mobile in UL and DL directions, which are given by defined mobile classes. Examples of multislot configurations are: 2 + 1 (2 slots in DL plus 1 in UL), 3 + 1, 4 + 2, etc.

2.2.2.2 Additional Network Elements

New network elements are added to the existing GSM architecture, creating a system that is based on packet-switched technology. The new elements are practically a parallel transport network for the packet information, which reuses the air interface layer (BTSs) and some of the units in the core network (Figure 2.6).

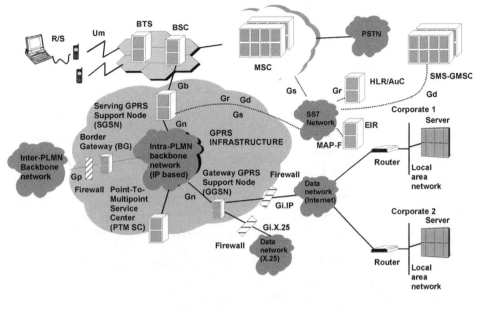

Figure 2.6   Functional view of GPRS architecture

The Packet Control Unit (PCU) is an extension to the BSC that implements the RLC/MAC procedures in the network side. It, in fact, contains the implementation of the different RRM algorithms. The PCU processes the data packets between the BTS and the SGSN, which are coming through the Abis and the Gb interfaces, respectively.

The Serving GPRS Support Node (SGSN) is a switching element that can handle communications with several PCUs associated to different BSCs. It takes care of the mobility management and authentication for GPRS users, and has register functionalities. The SGSN performs the adaptation between the IP world and the GPRS protocol stack. IP packets are received from the Gateway GPRS Support Node (GGSN) and converted into LLC data units in order to be transmitted to the BSS subsystem. Protocol conversion is done through the SNDCP layer. The communication with the PCU is performed through the Gb interface (Iu in later GERAN releases) and with the GGSN via the Gn interface.

The GGSN provides connectivity to external packet networks for the GPRS connections. It is linked to the different SGSNs through the Gn interface, and to the external packet networks through the Gi interface.

2.2.2.3 GPRS Protocols and Procedures

The treatment of the data and the new procedures conceived represent a major change compared to the voice system. The introduction of new coding schemes in the physical layer, the new RLC/MAC protocols and the RRM functionalities, such as scheduler, link adaptation, cell reselection or power control has a special relevance, due to their direct impact on service performance (Figure 2.7).

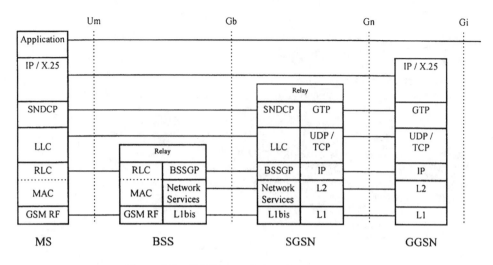

Figure 2.7   GPRS transmission plan protocol stack

The most remarkable aspect of the Medium Access Control (MAC) protocol is the multiplexing schemes used, where different users can share the same physical resources (PDCHs) by means of time multiplexing. The maximum number of simultaneous users in 1 TSL is restricted to 16 by system limitations. In case of UL transmission, the limitation is up to 7 simultaneous users.

The Radio Link Control (RLC) protocol includes mechanisms for error correction based on the retransmission of erroneous blocks, thus providing a reliable link to upper layers. It defines several transmission modes: acknowledged, unacknowledged and transparent.

Embedded in the new standard is a set of new concepts applicable to packet data calls and that are also important to understand the behavior and performance of the system.

- *GPRS attach/detach*: This procedure is used to register the mobile in the GPRS network, and activate a mobility management context for that user. Under this state no data can be exchanged through the GPRS channels until a PDP context has been created.

- *Packet Data Protocol (PDP) context*: Defines the context for a data call, including the Quality of Service (QoS), address, etc. A PDP context needs to be created before any data transaction is performed, and several PDP contexts can be created for one mobile at the same time. Unless the mobile is on 'always-on' mode, every time a new GPRS transaction needs to be performed there will be a delay associated to the establishment of the PDP context.

- *Temporary Block Flow (TBF)*: Data transfers within GPRS are made through TBF, a temporal connection between the MS and the PCU to transmit information in a specific direction. There are independent TBFs for UL and DL transmission, although both can coexist at the same time. All TBFs are released when the data transfer is finished at the LLC level, and needs to be opened again when new data arrives. Additional delays caused by this procedure have been shown to have an impact on service performance (section 6.4), especially with bursty data. In order to minimize this effect, special procedures to keep the TBF alive for a certain time after the data transmission were defined in the standard.

- *Quality of Service (QoS)*: The GPRS standard includes support for services differentiation based on certain parameters that can have an effect on the throughput and delay for the different connections, although the implementation of the mechanisms to support QoS is vendor specific. Four different classes are defined in the standard: conversational, streaming, interactive and background (see Chapter 4 for further details).

- *Mobility management*: During an active data transfer, the serving cell may change without an interruption on the TBF. A cell change can be triggered at the mobile station (autonomous cell reselection) or at the network side, based on measurements sent from the terminal (network controlled cell reselection). Unlike GSM voice service, the cell reselection is a blind procedure, meaning that the mobile releases the resources in the serving cell before being allocated new resources in the target cell. The establishment of the resources in the new cell is the reason for an extra delay during a cell reselection.

2.2.3 EDGE: The GSM Evolution

Enhanced Data Rates for GSM Evolution (EDGE) is an enhancement of GSM/GPRS system that increases its capacity and improves its quality and speed. The most important change coming with EDGE is the improvement of the modulation used, from GMSK to 8PSK, which triples the data transfer speed in the air interface: from 22.8 to 68.4 kbps per TSL.

While applicable to voice and data services, it is more commonly applied to GPRS because of the qualitative difference in terms of supported data rates. EDGE as an extension to GPRS has been early adopted in the USA markets and its footprint is increasing more and more, gaining momentum in Latin America, Asia and in some European countries.

Focusing on the improvements on the GPRS network, apart from the change in the air interface bit rate, there are other enhancements that are worthwhile mentioning.

- New set of coding schemes (MCS1–9) that can take advantage of the new transmission conditions. The link adaptation algorithm will select at any moment the most adequate codec. In case of high block error rate, the erroneous blocks can be retransmitted with a more robust codec than the one chosen for the first transmission.
- Incremental redundancy (IR) is a hybrid ARQ procedure that can be applied to retransmitted blocks in order to improve the probability of correct reception. IR provides relevant gains to the performance, and its use is particularly important to channels with FH.
- Longer RLC transmission windows, which reduces the probability of stalling.

2.2.4 (E)GPRS Performance

The performance of the (E)GPRS connections is very much dependent on the radio conditions (C/I) and on the available resources for data calls (PDCH). The radio conditions will determine what coding scheme shall be used, as well as the number of retransmissions that will be caused due to bad radio reception. In addition, dimensioning of the network may also force users to share the same TSLs, and the available bandwidth, when traffic load grows.

In general terms, due to the poor gains achieved with FH in GPRS, and to the fact that BCCH band usually have better C/I conditions, this is normally the preferred option for

Table 2.1 Typical GPRS and EGPRS RLC throughput per MS class

Throughput (kbps)	Mobile MS class 2	Mobile MS class 6	Mobile MS class 10
Configuration	2 + 1	3 + 1 / 2 + 2	4 + 1 / 3 + 2
GPRS			
Maximum	40	60	80
Typical (CSI-2) good C/I	22	32	42
Typical (CSI-2) medium/low C/I	16 / 20	30 / 24	40
Typical (CSI-4) good C/I	30	45	60
Typical (CSI-4) medium/low C/I	22 / 30	30 / 45	4060
EGPRS			
Maximum	118.4	177	236.8
Typical good C/I	100	150	200
Typical medium/low C/I	64 / 100	98 / 150	130/200

packet data connections. However, when GPRS traffic volume increases other deployments will need to be considered. In any case, the competition between GSM voice and GPRS resources, and different interference impact of data traffic on voice quality has to be taken into consideration [1].

Table 2.1 summarizes the maximum and typical throughput values that can be achieved in GPRS and EGPRS networks, for different mobile classes typically available today.

2.3 WCDMA/HSDPA

Among the available 3G systems, the scope of this section is focused on the Wideband Code Division Multiple Access (WCDMA) system, that has been standardized in the 3GPP. This system constitutes a joint effort of the standardization bodies from Europe, Japan, Korea, USA and China. The WCDMA system standardized by 3GPP is called Universal Terrestrial Radio Access (UTRA), and has two modalities: FDD and Time Division Duplex (TDD). In this section, the UTRA FDD system is considered for further discussion. For simplicity, in the sequel this system will be referred to as UMTS (Universal Mobile Telecommunications System), for which a comprehensive description can be found in [2].

This section is a short description of the basic architecture and features of UMTS (meaning UTRA FDD). The description is in line with the specifications of Release '99 and the most relevant upgrades for Release 5 with High-Speed Downlink Packet Access (HSDPA). Note that this description is not meant to be exhaustive, but to provide the reader with an insight into the system that is enough to understand the system specific issues that arise throughout the rest of the book.

2.3.1 System Architecture and RRM

The UMTS system architecture is sketched in Figure 2.8, which shows the logical network elements, the names of the different interfaces and the manner in which the RRM algorithms are distributed among the network elements. In UMTS terminology, the term 'Node-B' refers

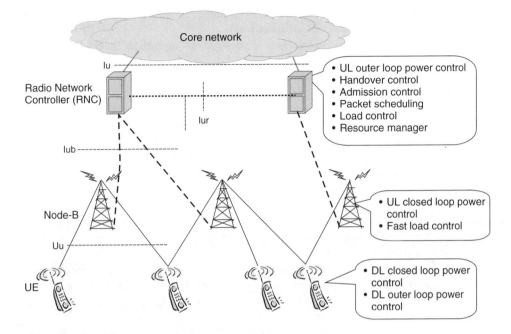

Figure 2.8   UMTS system architecture

to what is usually called base station, while the term 'user equipment' (UE) refers to the terminal. In the sequel, the UMTS terminology presented here will be used.

The Radio Network Controller (RNC) owns and controls the radio resources in its domain, which is formed by the set of Node-Bs connected to it, and it takes care of the following RRM algorithms: UL outer loop power control (OLPC), handover control (HC), packet scheduling (PS), admission control (AC), load control (LC) and resource manager (RM). When a UE is connected to the UMTS network (i.e. when it is not in idle mode), its identity is known by its serving RNC.[2] Moreover, the RNC interacts with the core network and also terminates the Radio Resource Control (RRC) protocol that defines the messages and procedures between the UE and the UMTS network [3]. In fact, most of the control signaling between the UE and the network is based on RRC messages, which carry all the parameters that are needed to set up, modify and release connections. For example, all the measurement reports and commands related to UE mobility across the network are conveyed by means of RRC signaling.

The Node-B converts the data flow between the Iub and the Uu interfaces [2], and it is responsible for fast UL closed-loop power control (CLPC) and fast LC. Furthermore, the UE is responsible for fast DL CLPC and DL OLPC. As shown in Figure 2.8, a UE can be connected to several Node-Bs via soft handover (SHO), no matter which RNC each Node-B is connected to. Furthermore, RRM algorithms can be classified into cell specific and

---

[2] For simplicity, the possibility of having a UE connected to several RNCs is not considered in this preliminary discussion. Thus, further clarification of concepts such as *serving* and *drift* RNC is not required. For further information in this respect, see [34].

connection specific. The connection specific RRM algorithms are: UL/DL OLPC, UL/DL CLPC and HC. The rest of them are cell specific.

In the following, the working principles of the main RRM algorithms are described [2, 4].

- *Fast CLPC* – is responsible for controlling the transmit power at each link via a closed-loop feedback scheme, in order to fulfill the SINR target in all the links while minimizing the total amount of transmitted power. In UL, such algorithm is essential in order to overcome the so-called near–far effect [5], and its use in DL increases the capacity [6].
- *OLPC* – adjusts the SINR target for CLPC in order to maintain the quality of the communication in terms of frame erasure rate (FER) at the desired level.
- *HC* – is needed in order to provide mobility across the network, supporting robust transitions between cells. HC decides the set of Node-Bs the UE should be connected to. Such decision can be based on coverage and/or load reasons, and is normally aided by pilot quality measurements conducted at the UE. In UMTS, a UE can be connected to several Node-Bs at the same time via SHO, which provides macro and micro diversity protection and guarantees a smooth transition between cells.
- *AC* – is responsible for controlling the load of the system so that the available capacity can be exploited without compromising the system stability. Before admitting a new UE or modifying the connection of an already admitted UE, AC checks whether these actions will sacrifice the planned coverage area or the quality of the existing connections. When a new UE is admitted or an existing connection is modified, AC is also in charge of setting the parameters for the new connection, e.g. the initial DL transmission power.
- *PS* – is the algorithm in charge of coordinating the resource allocation for non-real-time (NRT) traffic. The QoS requirements of the different UEs must be fulfilled while making an efficient use of scarce resources, so that the system capacity is maximized under the given constraints.
- *RM* – is the name of the algorithm that coordinates the distribution of the code resources among the different UEs in an efficient manner.
- *LC* – makes sure that the system is not overloaded, so that stability is not compromised. Basically, when an overload situation occurs, LC must bring the load back to the targeted levels. In order to do so the possible actions are: inter-frequency or inter-system handover for some UEs, quality decrease for some connections, throughput decrease for packet traffic and controlled dropping of low-priority UEs. These actions are taken at the RNC. However, other related actions can be taken at the Node-B by means of the so-called fast LC algorithm.

2.3.2 Transport Channels and their Mapping to the Physical Layer

Transport channels are, in general, services that are offered by Layer 1 to the higher layers in order to convey the data generated at these layers. Such channels are mapped onto different physical channels, which are defined by a specific carrier frequency, scrambling code, channelization code, start and stop time (giving a duration) and, on the UL, relative phase (0 or $\pi/2$) [7].

In Release '99, there are two types of transport channels: dedicated transport channels and common transport channels [7]. The only dedicated transport channel is the dedicated channel (DCH). There are six types of common transport channels: broadcast channel (BCH), forward access channel (FACH), paging channel (PCH), random access channel (RACH), UL common packet channel (CPCH) and downlink shared channel (DSCH). Further explanations for each

Transport channels   Physical channels

DCH   —————— Dedicated physical data channel (DPDCH)

                            Dedicated physical control channel (DPCCH)

RACH   —————— Physical random access channel (PRACH)

CPCH   —————— Physical common packet channel (PCPCH)

                            Common pilot channel (CPICH)

BCH   —————— Primary common control physical channel (P-CCPCH)

FACH   ————\   Secondary common control physical channel (S-CCPCH)

PCH   ———__—

                            Synchronisation channel (SCH)

DSCH   —————— Physical downlink shared channel (PDSCH)

                            Acquisition indicator channel (AICH)

                            Access preamble acquisition indicator channel (AP-AICH)

                            Paging indicator channel (PICH)

                            CPCH status indicator channel (CSICH)

                            Collision-detection/channel-assignment indicator

                            Channel (CD/CA-ICH)

HS-DSCH   —————— High-speed physical downlink shared channel (HS-PDSCH)

                            HS-DSCH-related shared control channel (HS-SCCH)

                            Dedicated physical control channel (UL) for HS-DSCH (HS-DPCCH)

Figure 2.9   Mapping of transport channels onto physical channels (Release '99) [7]

channel can be found in [2], and detailed information about how these channels are mapped onto the physical layer is given in [7] and illustrated in Figure 2.9.

One DCH is exclusively allocated to one UE. The DCH conveys all the information intended for that UE coming from higher layers, including data for the actual service. It can be used for both UL and DL, and supports CLPC with one power update per slot,[3] multi-code operation, bit rate variations with one frame resolution, SHO and the use of adaptive antennas. With Spreading Factor (SF) 4 and three parallel multi-codes, a DL DCH can carry approximately 2.8 Mbps when the coding rate equals ½ [7]. The same approximate bit rate can be obtained for the UL with SF 4, three parallel multi-codes and a coding rate of ½.

All Release '99 transport channels are terminated at the RNC. Thus, retransmissions for data packets are controlled by the RLC functionality at the RNC.

2.3.3 Physical Layer and Air Interface

UMTS uses WCDMA as a multiple access technique. WCDMA is a direct-sequence (DS) Code Division Multiple Access (CDMA) technique that consists in spreading the information bits over a large (wide) bandwidth by means of multiplying them with the pseudo-random

---

[3] The slot duration is 0.667 ms, and the radio frame duration is 10 ms, i.e. 15 slots [7].

bits (chips) of the spreading code [2]. In order to support very high variability of the bit rates, the use of a variable SF and several parallel channelization codes in the same connection is supported. The basic principles of CDMA are described in [8] and [9].

The chip rate is 3.84×10^6 chips per second, with a carrier bandwidth of 5 MHz. The large bandwidth allows the support of high user data rates, and opens for exploitation of multipath diversity. Separate 5-MHz frequency bands are used for UL and DL.

In DL, signals transmitted at the same cell are separated by means of synchronized orthogonal codes (referred to as channelization codes) extracted from an orthogonal variable spreading factor (OVSF) code tree [10], which is derived from the set of Walsh codes. Since orthogonal codes do not have white-noise properties, the total transmitted signal at each cell is scrambled by a pseudo-noise (PN) sequence that is referred to as scrambling code. The scrambling codes are complex-valued, and are obtained by I-Q multiplexing a Gold code and a delayed replica of the same Gold code. According to the UMTS specifications, only one OVSF code tree per scrambling code is available [11], which imposes a hard limit on the cell capacity that can be achieved with one single scrambling code per cell. In radio channels with no time dispersion, signals transmitted under the same scrambling code are fully orthogonal. However, this orthogonality is partly destroyed in time dispersive radio channels, and the part of the interference that is not orthogonal is just attenuated with the processing gain when despreading the desired signal [12]. The processing gain is defined as the ratio between the chip rate and the bit rate. Note that the DL DCH is mapped onto a dedicated physical data channel (DPDCH) and a dedicated physical control channel (DPCCH) (Figure 2.9), which are time multiplexed forming a DL dedicated physical channel (DPCH) [7] (Figure 2.10).

Figure 2.11 illustrates how several DL DPCHs are transmitted under the same scrambling code at the Node-B. The modulation scheme for the DL DPCH is quadrature phase shift keying (QPSK).

In UL, a scrambling code per UE is used. Channelization codes are used to separate different channels that are transmitted by a certain UE at the same time. Signals transmitted from different UEs use different scrambling codes. Thus, when despreading the signal from one UE, the noise and the signals coming from other UEs are attenuated with the processing gain. In UL, a DCH is mapped onto a DPCCH and one or more (up to 6) DPDCHs, which are I-Q/ code multiplexed (Figures 2.12 and 2.13). The modulation scheme for the UL DPDCH and the UL DPCCH is binary phase shift keying (BPSK).

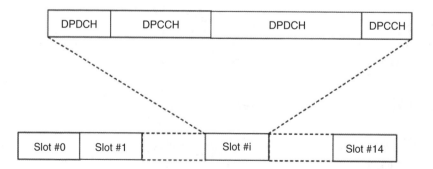

Figure 2.10   Time multiplexing of a DPDCH and a DPCCH in order to form a DL DPCH [7]

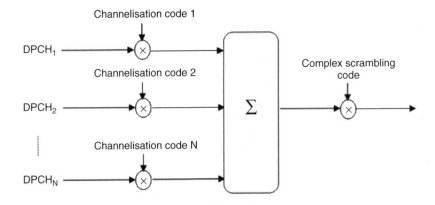

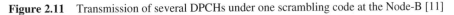

Channelisation code 1

DPCH$_1$

Channelisation code 2

DPCH$_2$

Channelisation code N

DPCH$_N$

Σ

Complex scrambling
code

Figure 2.11   Transmission of several DPCHs under one scrambling code at the Node-B [11]

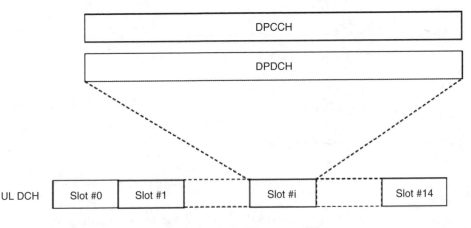

DPCCH

DPDCH

UL DCH | Slot #0 | Slot #1 | | Slot #i | | Slot #14

Figure 2.12   UL DCH structure [7]

2.3.4 The HSDPA Concept

In HSDPA, a new transport channel called the high-speed downlink shared channel (HS-DSCH) is introduced in order to achieve higher data rates. The HS-DSCH can be time and/or code multiplexed between the UEs in a cell, and is transmitted with fixed SF. All the UEs with access to the HS-DSCH have an associated downlink DPCH. The HS-DSCH is mapped onto one or several high-speed physical downlink shared channels (HS-PDSCHs), which can use QPSK or 16 QAM (Quadrature Amplitude Modulation). Note as reference that the DPCH can only use QPSK for DL transmission. Moreover, two other physical channels are included to facilitate the HSDPA operation: (i) the high-speed shared control channel (HS-SCCH), which carries the key information necessary for HS-DSCH demodulation; and (ii) the DL high-speed dedicated physical control channel (HS-DPCCH), which carries the ACK/NACK (acknowledge/negative acknowledgement) messages and the Channel Quality Indicator (CQI) feedback from the UE to the Node-B.

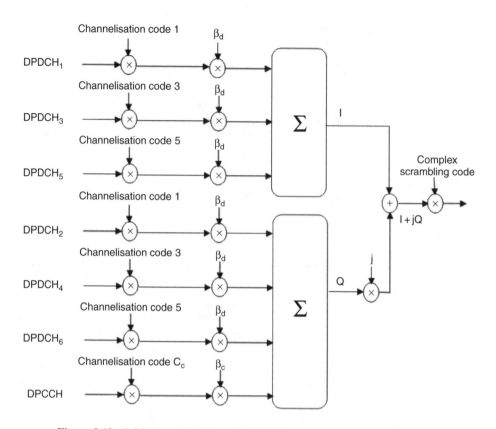

Figure 2.13 I-Q/code multiplexing of a DPCCH and several DPDCHs at the UE

Given certain transmit power for the HS-DSCH, the bit rate is adapted to the state of the radio channel by varying the Modulation and Coding Scheme (MCS) and the number of used HS-PDSCHs according to the channel quality estimates fed back from the UE. This adaptive variation of the transmission parameters is referred to as Link Adaptation (LA), and can be done every 2 ms (i.e. three slots), which is the duration of the radio frame, also referred to as Transmission Time Interval (TTI). The channel quality estimates for the LA algorithm can be obtained from the CLPC commands for the associated DL DPCH, the ACK/NACK ratio or the CQI sent by the UEs through the UL HS-DPCCH [13]. Note that fast power control is not allowed on the HS-PDSCHs.

A fast physical layer retransmission scheme with chase combining or incremental redundancy (Hybrid Automatic Repeat reQuest, H-ARQ) is specified which provides time diversity and facilitates faster retransmissions than in Release '99, where this process is handled by the RLC functionality at the RNC. In addition, the PS functionality is also moved from the RNC to the Node-B, which enables the possibility to consider the fast variations of the radio channel in the scheduling decisions, which can be made on a TTI basis.

2.4 IS-95/CDMA2000-1x, EV-DV, EV-DO

CDMA2000 as described in TIA/EIA/IS-2000 [14] is referred to as the 3G evolution of the 1.25 MHz 2G IS-95 [15] CDMA cellular radio standard, which was mostly pushed by

Qualcomm in the early stages. The IS-95 (or cdmaOne) standard was developed to overcome the capacity problems of the existing 800-MHz analog network. The first aggressive capacity promises by CDMA proponents were based on very approximate analysis and suggested that CDMA could offer capacity of up to 40 times that of 7-cell 30-kHz analog cellular. The first IS-95 version referred as IS-95A was standardized in 1995, although the first practical implementations were delayed until the end of 1996. The system was mainly targeted for voice services, although circuit-switched data connections at 14.4 kbps are also possible.

The first enhancement of the standard, IS-95B (1997), mostly improved the data capabilities of the IS-95A, introducing a new supplemental channel (SCH). The traffic channel was divided into one fundamental channel (FCH) and up to seven SCHs, which in theory provided maximum bitrate of 8×14.4 kbps $= 115.2$ kbps compared to the maximum 14.4 kbps of the IS-95A. The peak rate of any commercial system was however limited to 64 kbps. The world's first commercial launch of the IS-95B network was by Korea Telecom Freetel in July 1999.

CDMA2000-1x Revision 0, which was standardized in 2000, provided the possibility to use multiple code channels combined with multiple channel coding rates to achieve higher data rates and an adaptive link. Fast DL power control and introduction of QPSK modulation provided significant enhancements to voice capacity compared to IS-95. Compared to 14.4-kbps circuit-switched data connection of IS-95, 1x is able to provide data rates up to 153.6 kbps in Layer 1 in the SCH. CDMA2000-1x Revision A standardized in 2001 provided the signaling support for the so-called 3x (multicarrier 3×1.25 MHz) and provided theoretical peak bit rate of 1.037 Mbps.

1xEV-DO (1xEvolution Data Only), also known as or HDR (High Data Rates) due to naming of Motorola's 1xEV-DO approach, was a new approach to enhance the data capabilities of CDMA 1x system. In 1xEV-DO the 1.25-MHz carrier is dedicated for data traffic only and thus this is normally overlaid over voice capable 1x network. The peak data rates up to 2.46 Mbps and 153.46 kbps in forward (DL) and reverse (UL) links respectively are achieved by introducing advanced but very fragile modulation techniques with associated link adaptation to adapt the modulation to the channel conditions. The first 1xEV-DO network was launched by SK Telecom (Korea) in January 2001. This first revision (Revision 0) of EV-DO has also an enhanced Revision A which can provide 3.09 Mbps and 1.8 Mbps in forward and reverse links respectively. This standard also introduced QoS and multicasting capabilities to CDMA2000-1x.

1xEV-DV (1xEvolution Data and Voice) was the next evolution step that introduced the possibility for simultaneous voice and data. The first version of 1xEV-DV, CDMA2000 Revision C, was approved in 2002. The standard provided high-speed data and voice simultaneously, with peak bit rates of 3.09 Mbps/614.4 kbps in forward and reverse links respectively. The new standard was using adaptive coding and modulation where forward packet data channel (F-PDCH) carries variable packet size using multiple coding and modulation scheme (MCM). Each F-DPCH packet can be of 1, 2 or 4 slots (slot = 1.25 ms) and is modulated using QPSK, 8 PSK or 16 QAM. The rate control of forward link is based on channel conditions reported every 1.25 ms by mobile. Revision D of the EV-DV standard provided enhancements to reverse link (reverse packet data channel, R-PDCH) increasing the theoretical maximum bit rate of the reverse link to 1.5 Mbps. The reverse link has different modes of operation, including autonomous, rate control or scheduled mode operation. The high system throughput and low service-latency improvements are based on 5-ms R-PDCH frame, 3 HARQ channels and Autonomous Adaptive Incremental Redundancy (AAIR). In general, the standard is very similar to HSDPA with narrow bandwidth.

Today, 91 operators have launched 88 CDMA2000 1x and 11 1xEV-DO commercial networks across Asia, the Americas and Europe. As of 10 September 2004, 18 1x and 18 1xEV-DO networks were scheduled to be deployed in 2004 [16]. CDMA2000-1xEV-DV has not had practical commercial implementations at the time of writing this book.

2.4.1 CDMA2000-1x vs 3GPP UMTS

On high level, the system architecture of a CDMA2000 network is fairly similar to the 3GPP-based UTRA FDD network architecture. The radio access network architecture of CDMA2000 is, however, not strictly specified and thus, depending on the vendor the level of functionality in the base station will vary from vendor to vendor and in some implementations the base station can be connected even directly to the MSC while in some others there exists a Radio Network Controller (RNC) type of entity. In general, there are no open interfaces between different network elements as in WCDMA.

The network design of CDMA2000 network is based on the use of synchronized base stations. Terminals have been designed accordingly to search for the different base stations with identical timing but different code phase. The timing element needs to be therefore accounted for in the network planning process.

The key differences of CDMA2000-1x and WCDMA network are listed in Table 2.2. One of the main points, in addition to the channel bandwidth and chip rate is the synchronous mode of operation of a CDMA2000-1x network. Also, Table 2.3 shows the main technical enablers of HSDPA and 1xEV-DV which are the respective high-speed data evolutions for WCDMA and CDMA2000-1x respectively.

2.4.2 CDMA2000-1x Reference Architecture and QoS

Figure 2.14 depicts the reference model and Figure 2.15 the protocol reference model of CDMA2000-1x simple IP access as defined by 3GPP2. At the IP layer, the MS is attached

Table 2.2 Main differences of 3GPP WCDMA and 3GPP2 CDMA2000-1x networks

	3GPP WCDMA	3GPP2 CDMA2000-1x
Carrier spacing	5 MHz	1.25 MHz
Chip rate	3.84 Mcps	1.2288 Mcps
Power control frequency	1500 Hz, both in uplink and downlink	800 Hz, both in UL and DL
Frame length	10 ms with 15 slots	5, 10, 20, 40, 80 ms
Channelization code	OVSF codes	Walsh codes (reverse link) Walsh codes or quasi-orthogonal codes (forward link)
Inter-base station operation	Asynchronous	
Synchronous (optional)	Synchronous (typically obtained via GPS)	
Inter-frequency handovers	Yes, measurements with slotted mode possible, but measurement method not specified	

Table 2.3 Comparison of main technical characteristics of HSDPA vs 1xEV-DV

Characteristics	HSDPA	1xEV-DV
Forward link frame size	2 ms TTI (3 slots)	1.25, 2.5, 5, 10 ms Variable Frame Size (1.25 ms Slot size)
Channel quality reporting	Quality reporting rate 500 Hz (every 2 ms)	C/I reporting rate 800 Hz (every 1.25 ms)
Data user multiplexing	TDM/CDM	TDM/CDM (variable frame)
Adaptive modulation and coding	QPSK and 16 QAM mandatory	QPSK, 8-PSK and 16 QAM
Hybrid ARQ	Chase or IR	Asynchronous IR
SF	SF = 16 using UTRA OVSF channelization codes	Walsh code length = 32
Control channel approach	Dedicated channel pointing to shared channel	Common control channel

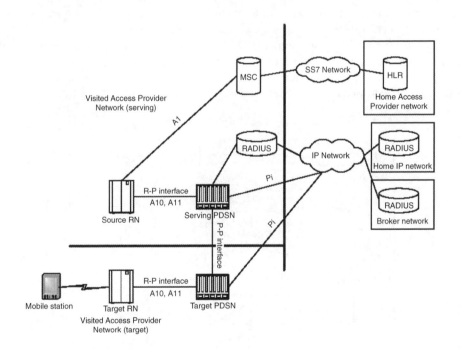

Figure 2.14 Reference model for simple IP access [18]

directly to the PDSN, as shown in the protocol stack in Figure 2.14. QoS is enabled at the Radio Link Protocol (RLP) and MAC layers over the air interface, which is the bottleneck in terms of resources in this architecture. There are several ways in which the MS acquires its IP address from the network. Independent of the acquisition of the IP address, the MS and the Packet Data Support Node (PDSN) establish a PPP session which is composed of one main, and several auxiliary service instances.

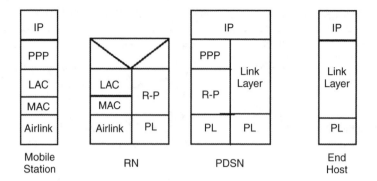

Figure 2.15 Protocol reference model for simple IP [19]

In current specifications, the QoS parameters are mapped onto the different service instances. Figure 2.16 describes the relationship between the PPP session and the services instances. The current specifications support up to six multiple service instances per MS, each of which may have associated RLP and/or QoS parameter settings.

A service instance may carry multiple flows. A flow is a series of packets that share a specific instantiation of IETF protocol layers. For example, an RTP flow may consist of the packets of an RTP/UDP/IP protocol instantiation, all of which share the same source and destination IP addresses and UDP port numbers.

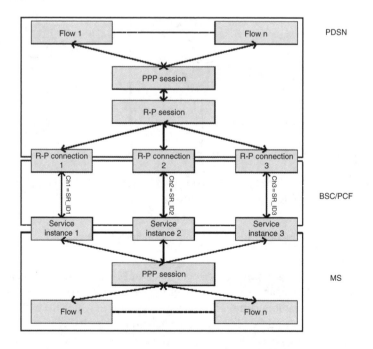

Figure 2.16 PPP session and multiple service instances [18]

When the MS establishes a packet data service, it shall originate a main service instance of Service Option (SO) type 33, for PPP negotiation before originating other auxiliary service instances [17]. The SO 33 defines the procedures for the data service and call control operation for MS and BSC, including a set of QoS parameters. When SO 33 is established, FCH is used by default and F/R-SCH is established on need basis to enable higher data rates.

2.4.3 Basic Voice Service with CDMA2000

While the current CDMA2000 systems mostly use Enhanced Variable Rate Codec (EVRC), the Selectable Mode Vocoder (SMV) will provide significant capacity and quality gains on CDMA2000 systems. SMV algorithm development was completed in February 2001, and it has been released to the 3GPP2 community for implementation. Products based on the SMV technology are expected by the middle of 2005. SMV's advances come from improvements in speech-encoding technology. Similar to GSM AMR, SMV also implements an algorithm to continually choose optimal encoding rates based on the input-speech characteristics, thereby ensuring that the sound quality remains high.

SMV offers CDMA carriers the flexibility to trade-off between quality and system capacity. Wireless operators can gain up to 75% increase in system capacity compared to the current CDMA vocoders by using the lower-encoding rates of SMV. Wireless operators can also provide improvements in voice quality by using data rates similar to the current CDMA vocoders. SMV operational mode can be controlled on a static or dynamic basis, allowing carriers further efficiency in service at peak loaded times [20]. The four different speech encoding rates of SMV are:

1. Mode 0 – 8.12 kbps
2. Mode 1 – 5.79 kbps
3. Mode 2 – 4.44 kbps
4. Mode 3 – 3.95 kbps.

Figure 2.17 depicts the trade-off between quality and capacity using the SMV.

2.4.4 Packet Data Operation with CDMA2000-1x

The basic concept of CDMA2000 packet data operation with data rates above 14.4 kbps relies on the concept of using a fundamental channel (F-FCH) and supplemental channel (F-SCH). These channels have equivalent counterparts in WCDMA; DCH and DSCH, respectively. F-FCH is equivalent to functionality traffic channel (TCH) for IS-95. It can support data, voice, or signaling multiplexed with one another at any rate from 750 bps to 14.4 kbps. F-SCH supports high-rate data services, and network may schedule transmission on a frame-by-frame basis, if desired. In addition to F-FCH and F-SCH, the dedicated control channel (F-DCCH) can be used for signaling or bursty data sessions. This channel allows for sending the signaling information without any impact on the parallel data stream.

The reverse TCH structure is similar to the forward TCH. It may include R-PICH, a fundamental channel (R-FCH), and/or a dedicated control channel (R-DCCH), and one or several supplemental channels (R-SCH). Their functionality and encoding structure is the same as of

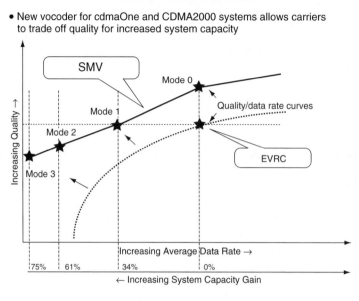

Figure 2.17   SMV quality vs capacity trade-off [16]

the forward link with data rates ranging from 1 kbps to 1 Mbps. (It is important to note that while the standard supports a maximum data rate of 1 Mbps, existing products are supporting a peak data rate of 307 kbps.)

The TCH structure and frame format is very flexible. In order to limit the signaling load that would be associated with a full frame format parameter negotiation, CDMA2000 specifies a set of channel configurations which define spreading rate and associated set of frames for each configuration.

The forward TCH always includes either a FCH or a DCCH. The existence of the SCH carrying the packet data is informed by higher layer signaling on the SCH. The structure also allows different handover configurations for different channels. For example, the F-DCCH, which carries critical signaling information, may be in SHO, while the associated F-SCH operation could be based on a best cell strategy.

Due to the bursty nature of packet data, it is very inefficient to dedicate a permanent TCH to a packet data call. SCHs can be assigned and de-assigned at any time by the base station. The SCH has the additional benefit of improved modulation, coding and power control schemes. This allows a single SCH to provide a data rate of up to 16 FCH in CDMA2000 Release 0 (or 153.6 kbps for Rate Set 1 rates), and up to 32 FCH in CDMA2000 Release A (or 307.2 kbps for Rate Set 1 rates). Note that each sector of a base station may transmit multiple SCHs simultaneously if it has sufficient transmit power and Walsh codes. The CDMA2000 standard limits the number of SCHs a mobile station can support simultaneously to two. This is in addition to the FCH or DCCH, which are set up for the entire duration of the call since they are used to carry signaling and control frames as well as data. Two approaches are possible: individually assigned SCHs, with either finite or infinite assignments, or shared SCHs with infinite assignments. The finite assignment is the usual approach since infinite assignment will waste the radio resources when there is no data to be sent.

The SCH allocation is typically based on the amount of data in the users RLP buffer (Figure 2.18). When the data waiting in the RLP buffer exceeds a threshold, PSDN/PCF asks the BTS for an F-SCH. The data rate of the F-SCH can be further limited by the available BTS forward TX power, available Walsh codes, multiplexing with other users waiting for F-SCH and mobile capability. The SCH allocation strategy may also be used to provide QoS differentiation between different services or traffic types.

When the RLP buffer is nearly empty, the SCH is released and only FCH is kept active. Different timers and/or hard-coded triggers may be applied to control the packet data state of the mobile. Figure 2.19 depicts an example of channel allocation during a Web browsing session where the FCH and SCH are allocated dynamically. The so-called dormant mode refers to CDMA2000-1x packet data state where the PPP connection between the mobile terminal and the Packet Data Support Node (PDSN) as well as R-P connection between Radio Network (RN) and PDSN are kept open but the packet SO is disconnected. The dedicated data channel can be reactivated when the data in the buffer again exceeds the defined threshold.

For bursty and delay-tolerant traffic, assigning a few scheduled fat pipes is preferable to dedicating many thin or slow pipes. The fat-pipe approach exploits variations in the channel conditions of different users to maximize sector throughput. The more sensitive the traffic becomes to delay, such as voice, the more appropriate the dedicated TCH approach becomes. In general, the CDMA2000-1x resource sharing between the different voice and data connections is fairly similar to the WCDMA, excluding the 1xEV-DO which can only carry data. Figure 2.20 depicts an example of dynamic resource sharing of CDMA2000-1x.

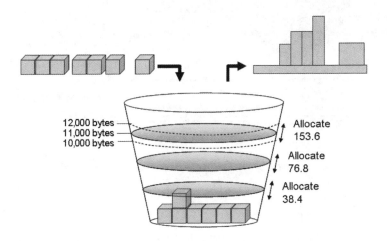

Figure 2.18 SCH rate allocation based on the amount of data in the users RLP buffer

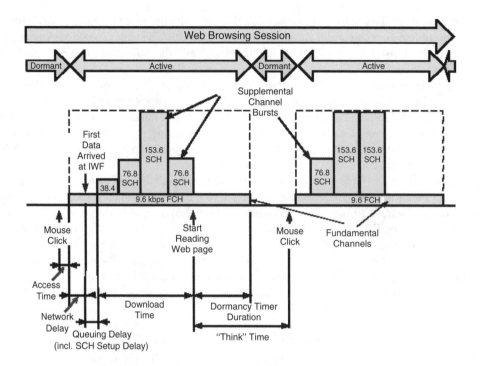

Figure 2.19   Channel allocation example – Web browsing session

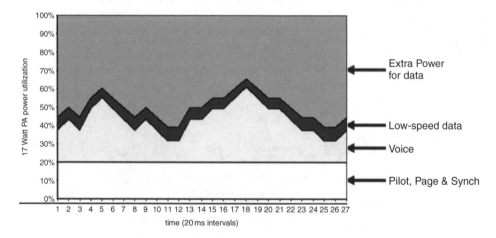

Figure 2.20   Resource sharing on one carrier (not applicable to 1xEV-DO)

2.4.5 CDMA2000-1x Performance

Obtaining the theoretical maximum throughput of CDMA2000-1x network is challenging due to the nature of radio channel and interaction of different protocols some of which were originally designed for fixed network. The TCP and application throughput are easily impacted by:

- Sub-optimal radio conditions
- SCH assignment algorithm/sub-optimal parameterization
- RLP
- Transmission window size
- Header compression
- Combination of any of the above.

In case of moderate FER, RLP usually can recover the data and no TCP retransmission is needed. RLP retransmissions consume part of the bandwidth and thus reduce the total throughput. Under bad radio conditions the RLP will get exhausted due to excessive amount of NACKs which easily leads to TCP slow start and retransmissions. At very severe FER conditions, the TCP timeout can cause the whole transfer to be halted. To services other than TCP based, which do not perform retransmissions at transport level (i.e. streaming over UDP), this effect can cause great degradation of the user quality.

As mentioned earlier in this chapter, the normal approach is to use finite assignments of the SCH to avoid waste of radio resources. Finite SCH assignments are, however, fairly sensitive and can cause bandwidth oscillations and thus inefficient use of the channel. Another drawback of finite assignment is the overhead that this procedure causes due to setting up of SCH.

Normally, the recovery of data on upper layer is more expensive than on lower layer and thus the RLP protocol should normally perform this function. In general, there are two RLP modes: transparent and non-transparent modes. While the transparent RLP does not have retransmission, the non-transparent RLP has. Non-transparent RLP is used in most CDMA2000-1x applications to provide reliable transmission of the data. RLP is a NACK protocol, which means that if a data frame is sent successfully, no ACK is sent. The ACK is sent only when an RLP frame is found lost. RLP uses a Best Effort protocol to retransmits the lost frame for certain rounds. If the error is still not recovered after a number of rounds of retransmission, RLP will handover the error to upper layer and let the upper layer error correction mechanism to take care.

The NACK protocol of the RLP impacts on both the utilization of radio resources and the Round-Trip Time (RTT) of the connections. RLP uses 20 ms time frame, i.e. it sends and receives RLP frames every 20 ms.

As described earlier in this chapter, RLP SCHs can be allocated and de-allocated dynamically during the data transmission. CDMA2000-1x RLP has different Multiplexing (MUX) options. With each MUX option the RLP can have different SCHs. Depending on the selection of the MUX, RLP can thus have different bandwidths.

Very important characteristic of the TCP protocol is the inter-dependence of the RTT and the TCP throughput. In addition to throughput of the radio link, the RTT is also directly proportional to the amount of data in the RLP buffer, which is mainly controlled by TCP congestion window. The bigger the congestion window, the more data the system allows to RLP buffer. On the other hand, very small congestion window restricts the overall throughput.

Finally, the header compression of the IP traffic has proved to have significant impact on the end-user perceived throughput. Different mechanisms can be applied to reduce the TCP/IP/PPP overhead, however the performance of the header compression techniques decreases in bad radio-link conditions.

2.4.6 Mobility

The different handover types supported by CDMA2000-1x system are very similar to those
of 3GPP WCDMA. The hard handover is performed between cells using different frequencies,
while a SHO is made between cells that use the same frequency. In SHO, the FCH is maintained and the SCH may or may not be turned off depending on the BS implementation. In
hard handover, the SCH is always turned off. Additionally, if the SO is changing in the hard
handover (typical inter-generation handover) the packet data state has to be changed to
dormant mode and the connection has to be re-established. Depending on the mobility level
of the handover (Figure 2.21), for example, the QoS re-establishment procedure changes.
QoS re-establishment is handled at Layer 2.

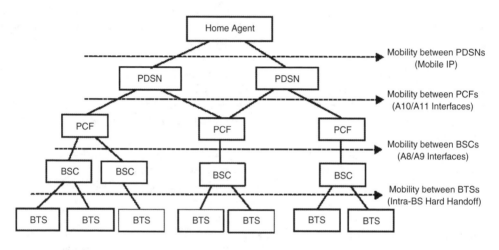

Figure 2.21   Conceptual view of packet data mobility levels in CDMA2000-1x network [21]

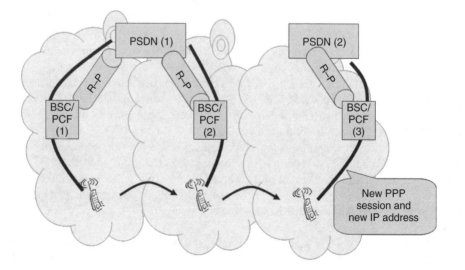

Figure 2.22   Example of mobility management in CDMA2000-1x network

There are two possible mobility modes in CDMA2000-1x network: simple IP and mobile IP. In simple IP, the IP addresses are assigned dynamically and the dynamic IP address is valid only in the PDSN coverage area. The mobile connects to external packet networks through the PDSN attached to the local BSC. The IP address for the connection is assigned by the local PDSN from the pool of IP addresses that are available for it. If the mobile moves to a different network, the data session ends and has to be restarted again in new network.

Mobile IP service can be based on static (public or private) or dynamically assigned IP addresses. With mobile IP, the handovers are possible between the different PDSNs. Mobile station is assigned a static IP address belonging to its Home Agent (HA). This means that mobile IP can be used to manage the mobility in the network between PDSNs (Figure 2.22).

2.5 WLAN

WLAN technology is a term used for a wide range of Wireless Local Access Network technologies. Those technologies aim to provide connectivity and wireless access at a high bandwidth to IP-based networks in a similar way or better than wired connections (e.g. Ethernet) provide nowadays. The different options that are currently available in the market have appeared accordingly to the progressive increase of higher bandwidths.

The first WLAN standard was created within the IEEE in 1997, the reference for this one is 802.11 (Table 2.4) [25]. The possibilities provided by this standard were to support a maximum of 2 Mbps using the unregulated radio-signaling frequency of 2.4 GHz. A drawback when using this unregulated radio-signaling frequency is that WLAN radio signals can be interfered by other equipment working in the same frequency range such as microwaves oven, cordless phones, etc.; in any case, by keeping these at a reasonable distance interference can be avoided. The benefit of using this unregulated radio frequency band is that the cost of the equipment can be lowered as there is no need to pay radio frequency licenses.

The 2 Mbps provided by 802.11 were appropriated but are too slow for lots of applications. This triggered the creation of a new IEEE standard, the 802.11b as an extension of 802.11. In September 1999, 802.11b was already supporting up to 11 Mbps, providing a bandwidth comparable to traditional Ethernet. 802.11b uses also the same radio frequency band (i.e. 2.4 GHz) as 802.11, having the same drawbacks and benefits derived from free spectrum, but providing a much more convenient bandwidth enough for the majority of applications.

Table 2.4   WLAN Standard version

IEEE 802.11	Standard for WLAN operations at data rates up to 2 Mbps in the 2.4 GHz ISM Industrial, Scientific and Medical (ISM) band.
IEEE 802.11a	Standard for WLAN operations at data rates up to 54 Mbps in the 5 GHz Unlicensed National Information Infrastructure (UNII) band.
IEEE 802.11b	Standard for WLAN operations at data rates up to 11 Mbps in the 2.4 GHz ISM band.
IEEE 802.11g	High-rate extension to 802.11b allowing for data rates up to 54 Mbps in the 2.4 GHz ISM band.

At the same time that 802.11b was being standardized in IEEE, another standard 802.11a was generated as an extension of the 802.11. The 802.11a standard was released in September 1999 supporting a bandwidth of 54 Mbps, providing a bandwidth more than enough for the majority of applications; actually, it provided enough bandwidth for several users at the same time. However, 802.11a needs the utilization of a higher frequency band to provide higher bandwidth, and it uses the 5 GHz radio frequency band, which is a regulated frequency band. The utilization of a higher frequency has several drawbacks: the achieved distance range is smaller compared to 802.11b, and also penetration of walls and obstacles is more difficult. Despite of providing higher bandwidth, 802.11a products came into the market later than 802.11b because the equipment was more expensive due to regulated band usage constrains. Commonly, 802.11b technology is used in domestic market and 802.11a is used in business market.

Between 2002 and 2003, a new standard called 802.11g was generated in IEEE. The 802.11g tries to combine the main advantages of 802.11a and 802.11b, so it is able to support a bandwidth up to 54 Mbps using the 2.4 GHz frequency band. The new standard was created to be backward compatible with 802.11b access points, so old devices can still work with new equipment, but at lower rates.

All the 802.11 technologies are commonly known as Wi-Fi (Wireless Fidelity). Wi-Fi Alliance is an entity that certifies that vendor's products follows the different 802.11 specifications. It certifies 802.11, 802.11a, 802.11b and 802.11g.

Bluetooth is another wireless network technology developed in a different path than 802.11 technologies. Bluetooth supports a very short range of approximately 10 meters, providing up to 1 Mbps. The most attractive point of Bluetooth is its low manufacturing cost, otherwise it is not a technology that can be considered for general-purpose networking due to the coverage and bandwidth restrictions. However, the range of application for controlling and messaging remote device is very wide. Nowadays it starts to be very common in cell phones or computers as a fast and easy way to connect with remote devices such as earphones, GPS, PDAs, etc.

2.5.1 Complementary WLAN Access Technology for Cellular Networks

The high data rates provided by WLAN technologies are very attractive for different purposes, and cellular network industry has started to put an eye on it as a good possibility to increase data rates provided to their customers. Maximum data rates provided by traditional cellular network technologies are poor compared to maximum data rates provided by any of the 802.11 families. As an example, the maximum throughput that can be achieved by an MS using legacy 2.5G EGPRS terminals is 59.2 kbps in single slot mode, or up to roughly 220 kbps with current 4 TSLs devices in best conditions. For WCDMA, this limit is set up to 384 kbps for current starting networks, although 2 Mbps would be also possible. This fact makes the WLAN technology very attractive for new upcoming 3G services that in general would require a high throughput. However, the main limitation is the coverage provided by WLAN equipment, which is remarkably shorter than the one provided by traditional cellular networks. This led to the conclusion that WLAN could be used to access upcoming new set of 3G services, but with a different scope as that of the so-called 3G cellular networks. Based on these assumptions, WLAN AN was also raised as a new complementary access technology to provide 3G services.

WLAN ANs are well suited to hotspot coverage, where there is a high density of high data rate services, such as 3G services, requiring a limited mobility (e.g. located in airports, cafeterias, etc.). But looking at the other side of the coin, 3G systems allow voice, support wide coverage area and provide high possibilities to mobility; 3G systems are more suitable for wider areas with relatively low to moderate demand of high data rate services but with more mobility needs. This implies that WLAN ANs and 3G systems may compete in certain market niche but more often the market niche for WLAN AN and 3G systems are complementary, which should enable a nice and soft interworking between these two types of technologies when accessing 3G services.

After understanding the complementary characteristics of WLAN and 3G cellular technologies, the next step is to provide multi-access functionalities to the terminals, and furthermore, to make multi-access solution work smoothly by providing seamless mobility mechanisms between WLAN and 3G systems.

2.5.2 WLAN-3GPP and WLAN-3GPP2 Architecture

The clear business opportunity created with WLAN ANs potential usage by 3G services triggered the technical development of standards in the different forums to enable a multi-vendor interaction between WLAN ANs, standardized in IEEE, and 3G cellular networks, mainly standardized in 3GPP and 3GPP2.

When providing interworking between WLAN AN and 3G cellular networks there are two main architectural approaches that can be taken: the 'tight-coupling approach' and the 'loose-coupling approach,' as depicted in Figure 2.23.

The *tight-coupling approach* is based on making the WLAN network appears to the 3G core as another 3G access network similar to RAN or BSS. The WLAN would need to emulate functions that are performed by already available 3G radio access networks. In this approach, the WLAN gateway hides to the 3G network the WLAN AN peculiarities and implements all 3G protocols such as mobility management, authentication, etc. The WLAN UE also needs to implement all the 3G protocols on top of standard WLAN protocols that will allow to inject 3G traffic to the 3G core. As conclusion, the tight-coupling architecture would be able to reuse authentication, signaling and billing infrastructure and protocols of the 3G cellular networks. However, this approach has some disadvantages since the 3G network is directly exposed to the WLAN AN. This has important implications from the security point of view, and would require that the cellular operator will have to own also the WLAN AN network, or subscribe complex agreements with third-party WLAN AN suppliers. In practice, this leads to the fact that third party's WLAN AN cannot be easily connected to the 3G-WLAN networks. On the other hand, the tight-coupling approach imposes a strong requirement in the WLAN UEs, since they should implement 3G protocol stack, leaving legacy WLAN UEs out of the game. The loose-coupling approach appears as an alternative to the concerns and problems that the tight-coupling approach imposes.

The *loose-coupling approach* is based on having the access to 3G services for WLAN UEs decoupled from native RAN technologies. Instead of connecting directly to internal 3G core switches (e.g. SGSN), the WLAN UE gets access directly to the 3G services via the Internet. However, even though WLAN UE data traffic is not injected in the 3G core network, the WLAN UE authentication, authorization to services and billing is performed by the 3G core operator in collaboration with the WLAN AN. There are two paths or two planes for user and

(a)

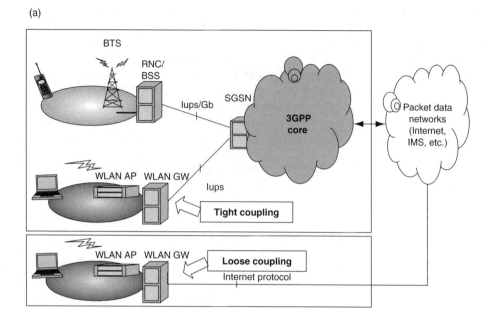

(b)

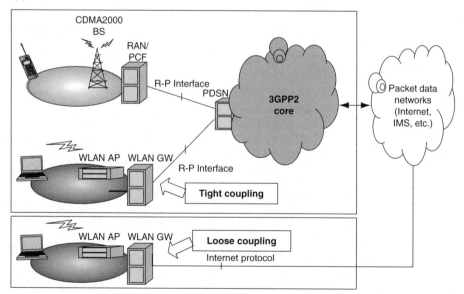

Figure 2.23   Tight coupling vs loose coupling for WlAN and 3GPP (a) and for WLAN and 3GPP2 (b)

control information. User plane, or WLAN UE data traffic, is generally carried from/to the WLAN AN to/from the Internet bypassing the 3G core network, whereas the control plane signaling is carried from the WLAN AN to the 3G core network and vice versa, in order to primary authenticate the user and authorize the utilization of the user plane path. This approach requires in the 3G core network to provision new equipment, mainly AAA servers, that are used for WLAN authentication/authorization and accounting.

One of the main drawbacks of the loose-coupling approach compared to the tight coupling is the fact of not allowing the reutilization of traditional cellular authentication and authorization infrastructure. However, it provides many other benefits that raise it as the future-proof solution for WLAN-3GPP/2 interworking.

The main benefits are that loose coupling imposes minimal changes in the WLAN AN, does not need the support of specific 3GPP/2 protocols in the WLAN UEs, using instead IETF-based protocols that are more familiar to the WLAN AN industry, and eases the access to IP-based services. Due to the decoupling of functionalities related to user plane and control plane, it also eases the path for future multi-access technologies (WLAN, xDSL, etc.) interworking solutions for 3GPP/2. The loose-coupling approach is the option followed by 3GPP and 3GPP2 in their standards for the WLAN-3GPP and 3GPP2 interworking, as explained in [22, 24] and [26].

The WLAN-3GPP standardization started in 3GPP during Release 6. Several WLAN-3GPP/2 scenarios were defined [23], in order to provide an incremental complexity of the interworking capabilities.

- *Scenario 1: Common billing and customer care*, This scenario targets to have a common bill and customer care for the user taking into account that two entities are involved: the WLAN AN provider and the cellular operator. This does not require any technical modification just a business agreement between the two parties (WLAN AN and cellular operator) to provide a common bill and customer care.
- *Scenario 2: 3GPP-system-based access control and charging*, In this scenario, authentication, authorization and accounting are provided by the 3GPP system, i.e. before the WLAN UE traffic data is able to flow, the 3GPP system has performed authentication/authorization of the user and the result has been signaled to the WLAN AN in order to grant or deny access to the WLAN UE traffic data.
- *Scenario 3: Access to 3GPP-PS-based services*, In this scenario, the target is to provide 3GPP-PS-based services to the WLAN UEs similarly to traditional cellular users. Those services are mainly IP-based services, such as pure Internet or more advanced type of services on the IMS.
- *Scenario 4: Service continuity*, In this scenario, the goal is to allow services supported in Scenario 3 to survive changes of access between WLAN and 3GPP systems. The change maybe noticeable by the user (short cut in the session, QoS degradation, etc.), but the user will not need to re-establish the service.
- *Scenario 5: Seamless services*, This scenario is a step ahead from Scenario 4, and targets to provide seamless changes of radio access methods.
- *Scenario 6: Access to 3GPP CS services*.

The current status of the standardization has not yet frozen for current release, but after the summer of 2004 everything seems to indicate that Scenario 2 and 3 will be included, by providing AAA authentication and authorization support, and access to Internet and IMS services. It is foreseen that Release 7 will include Scenario 4 support, most probably based on MIP. A similar development is happening in 3GPP2.

2.6 Future Outlook

Previous chapters presented an overview of different cellular technologies. The aim of this chapter is to provide a perspective on the evolution of the current mobile networks, although it is difficult to foresee which architectures and technologies will succeed in the forthcoming years.

The next step after standardization of the 3G infrastructures is often referred as Fourth Generation (4G), although the scope of the term is still vague. 4G may refer to a new radio interface standard, which will presumably provide higher data rates and better adaptation to the user requirements and service circumstances than 3G technologies. Some partial goals are easily anticipated, such as much higher spectrum efficiency and dynamic bandwidth allocation. After the success of Internet technologies and mobile communications, their combination can be considered one major agent toward 4G. In order to be really IP-enabled, 4G networks will require improvements in the architecture and the upper layers of the protocol apart from those in the stack physical layer.

2.6.1 Heterogeneous Networks

Already in the introduction section (Figure 2.1), several current and near-future access technologies classified by mobility and aggregated bit rate were shown. The variety of system architectures is manifest in the compatibility problems that this represents. In an effort to unify different technologies, the IEEE has established a hierarchy of complementary wireless standards, each one representing the optimized technology for a distinct usage model and designed to complement the others: IEEE 802.15 for Personal Area Networks (PAN) [29], IEEE 802.11 for WLAN [30], 802.16 for the Metropolitan Area Network (WMAN) [31, 32], and IEEE 802.20 for Mobile Broadband Wireless Access Network (MBWA) [33]. The works performed by 3GPP and 3GPP2 to merge WLAN with cellular systems were described in the previous section, but still the noticeable heterogeneity is unlikely to disappear in the foreseeable future after massive investments by operators in network infrastructure.

4G networks are expected to integrate all heterogeneous wired and wireless networks in order to meet the challenge of 'optimally connected anywhere, any time'. 4G networks are predicted to be based on a common, flexible and seamless all-IP protocol stack [34], where mobile terminals will need to be highly integrated multimode, multiband, and able to utilize a wide range of applications.

In order to provide the terminals with high reconfiguration capabilities, *Software Defined Radio* (SDR) gives the physical layer the flexibility to access different wireless systems by a single interface [35]. SDRs use adaptable software and flexible hardware platforms that enable reconfigurable system architectures for wireless networks and user terminals. These multi-functional terminals will enable users to communicate as they move between different wireless network environments.

However, even if the terminals can adapt to the different radio interfaces, maintaining the service continuity of mobile terminals through such diverse environment is a complex issue that the current relatively simplistic algorithms for handover cannot support. Handoff methods shall enable mobile terminals to maintain connectivity when moving between cells, between systems, between frequencies and at the upper layers, between IP subnets.

The problem of *vertical handover* [36] is defined in the context of heterogeneous network architectures: a handover for users that move between different types of networks. To perform handover decisions, the quality of the radio link channel is estimated to detect any signal degradation and to select the new cell to handoff. In heterogeneous networks, this means that the mobile terminal should be able to measure over technologically different systems while keeping the original link. Moreover, mobile nodes must be able to interoperate with different networks, and their corresponding signaling protocols, routing techniques and mobility management standards. On the other hand, sophisticated handover policies and strategies are able to exploit the increased available network resources; optimal decisions can be achieved by taking into account additional parameters related to factors such as system performance, network conditions, service type, and user status and preferences.

4G systems will provide access to rich content and multimedia applications, available even while the user is moving. To provide good end-user experience when using a particular service, mechanisms and protocols oriented to give different treatment to distinct data flows are necessary. Chapter 4 is devoted to these QoS mechanisms.

QoS features shall provide the required performance of latency, jitter and packet loss needed to support the service. QoS must be treated as an end-to-end issue, and handled by all the communication layers, since each layer may be required to provide a set of service guarantees. Two main factors influence QoS over wireless access networks [37]. First, a major problem is the coupling of QoS and mobility management, i.e. how to keep providing the same level of quality to the packet flow during and after a handover. Secondly, as mobile channel is time variant, it is necessary to efficiently manage the wireless resource dynamics. Adaptive modulation techniques can adapt the physical layer characteristics to the variant nature of the channel and adjust the used resources to the service needs. But adaptation strategies may also be applied to other network elements.

In order to adapt the system behavior to the changing communication context, cross-layer adaptation algorithms adjust the behavior of configurable components to the user, application, or system-performance requirements. These kind of algorithms are already present in legacy networks as adaptation of the transmission codec based on radio quality, or selection of proper transmission channel according to the bandwidth required by upper layers. Among others, some examples of cross-layer adaptation can be found in GSM: AMR codecs reduce the voice quality in order to provide higher protection to radio quality degradation; DTX is a method that momentarily powers-down a wireless terminal set when there is no voice input.

In the data side, schedulers' performance with access to signal and interference strength measurements (Channel State Information, CSI) is better than that of those which share resources with no knowledge of channel quality. In case that Service State Information (SSI) is also taken into account, the QoS fulfillment will be more easily accomplished. This tendency, already present in 2G and 3G as solution for specific problems, can be envisioned for the heterogeneous environments in 4G as intrinsic to the architecture design [38].

2.6.2 *Physical and MAC Layers Trends*

Due to the large variety of applications which are expected to be supported in 4G over a wide set of different environments, a unique optimal and universal solution for the physical and MAC layers does not seem feasible all over the forecast heterogeneous network.

2.6.2.1 Orthogonal Frequency Division Multiplexing (OFDM)

Current standardization works in ETSI and IEEE basically deal with two different modulation and multiple access schemes: CDMA and OFDM. In fact, both techniques compete to become the default in several standards already issued, e.g. WPAN IEEE 802.15. Some previous sections addressed CDMA-based 3G cellular system performance. Standards IEEE 802.16 [31, 32] (or ETSI HIPERMAN [39]), IEEE 802.11a [30] (or ETSI HIPERLAN type 2 [40]), and ETSI DVB-T [41] take advantage of multicarrier modulation to enhance their link capacity.

OFDM is a modulation technique used to counteract the effects of InterSymbol Interference (ISI) in frequency selective channels [42]. OFDM divides the transmission band into a large number of sub-bands narrow enough to be considered flat. The symbol sequence is split into lower-speed symbol streams transmitted simultaneously on the resulting comb of carriers. Often, an Inverse Fast Fourier Transform (IFFT) efficiently performs the modulation process. Its reciprocal process, the forward Fast Fourier Transform (FFT), can be used to recover the data if a cyclic extension of the OFDM symbol eliminates the residual ISI. In this way, OFDM can be considered as a time-frequency squared pattern, where each 'bin' can be addressed independently, as Figure 2.24 represents.

Modulation of the OFDM subcarriers is analogous to that of the conventional single carrier (SC) systems. The modulation schemes of the subcarriers are generally QAM in conjunction with convolutional coding. One major OFDM advantage is that each subcarrier in the symbol can be modulated with a different constellation-coding scheme, assigning high data rates to those carriers with good SNR and more robust schemes to carriers with low SNR. The bit allocation can be modified on a frame-by-frame basis, for example, to simultaneously track the time variant frequency response of the channel and fulfill the BER service requirements [43].

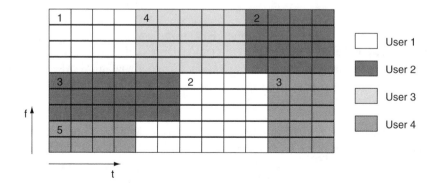

Figure 2.24   Example of OFDMA assignation. Tints distinguish users or services. The number on the top left of each data block stands for the number of bits per constellation used in that burst

When OFDM is also used as multiplexing technique, the term 'OFDM Access' (OFDMA) is preferred. In this case, a block of contiguous bins is assigned to a single user which can be considered a hybrid TDMA-FDMA technique. Figure 2.24 presents an example of channel allocation, where users are distinguished by different tints. The number of bits per constellation employed on each data block is adapted to the instantaneous channel response at that frequency band and to the current service state for that specific user.

High capacity and resource allocation flexibility are major advantages in OFDM. QoS requirements can be satisfied by the right selection of constellation-coding schemes and correct scheduling even over harsh environments [44]. Distinct users experience fading at different subcarrier. Allocating each user to those carriers with higher data rates maximizes the aggregated throughput and allows the system to handle a wide number of active users.

Obviously, OFDM also presents disadvantages, specially related to its high sensitivity to carrier offset [45]. This becomes even worse in mobile applications due to the time-variant nature of the mobile channel and the Doppler spread effect. In this case, the orthogonality between carriers is destroyed, which manifests itself in the form of Inter Carrier Interference (ICI) and results in an error floor worse as the symbol length increases [46].

2.6.2.2 Other Capacity Enhancement Techniques

Some standards as IEEE 802.11b or HIPERLAN type 2 assign the same constellation to all subcarriers in the OFDM symbol. Even in this case, multicarrier modulation is advantageous due to the frequency diversity benefit of the broad bandwidth.

A different kind of diversity, spatial diversity, can mitigate the effects of time and frequency fading [47]. In this technique, several antennas are placed at the receiver to ideally receive a set of uncorrelated signals. Those signals can then be processed by different methods (as Equal Gain Combining or Maximum Ratio Combining) to reduce the probability of losing the signal exponentially with the number of decorrelated antenna elements.

Another powerful use of antenna arrays is beamforming. In this case, the transmitter or the receiver (typically the BTS) is equipped with a set of antennas. The transmitter can weight the transmission on each antenna to modify the radiation pattern in order to focus the energy on the desired direction. In this way, the average signal level received by the target terminal is maximized while the interference power is minimized in other directions. Several recently developed standards, as 802.11a, propose this technique to increase the number of users simultaneously active. As a disadvantage, the terminal position has to be tracked during its movement through the cell.

Joint coding can be realized over multiple transmission antennas in a technique named Space-Time Coding (STC) [48]. In these schemes, a number of code symbols equal to the number of antennas in the transmitter are generated and sent simultaneously, one symbol from each antenna. These symbols are generated by the a Space-Time Encoder to maximize the diversity and/or the coding gain with the appropriate signal processing and decoding procedure.

As subscriber units are evolving to become sophisticated wireless Internet access devices, their stringent size and complexity constraints are relaxing somewhat. This makes multiple antenna element transceivers a possibility at both sides of the link. In this case, data are transmitted over a matrix rather than a vector channel, i.e. a Multiple Input Multiple Output (MIMO) channel [49]. 3GPP is currently working to integrate MIMO techniques into HSDPA channel. With the adequate signal processing, linear increase in the multiuser capacity is possible with no additional power or bandwidth expenditure. Although conditioned by the relevant implementation issues, MIMO techniques applied to OFDM open new and enormous opportunities beyond the added diversity or array gain benefits.

References

[1] T. Halonen, J. Romero and J. Melero, 'GSM, GPRS and EDGE Performance', Ed. Wiley, 2nd edition, 2003.

[2] H. Holma and A. Toskala (eds). 'WCDMA for UMTS. Radio Access for Third Generation Mobile Communications', 2nd edition, John Wiley & Sons Ltd, England, August 2002.

[3] 3rd Generation Partnership Project, 'Radio Resource Protocol (RRC) Specification', TS. 25.331, V5.2.0, Available online at www.3gpp.org, September 2002.

[4] 3rd Generation Partnership Project, 'Radio Resource Management Strategies', TR. 25.922, V5.0.0, Available online at www.3gpp.org, March 2002.

[5] C. Lee and R. Steele, 'Closed-loop Power Control in CDMA Systems', *IEE Proceedings on Communications*, Vol. 143, Issue 4, August 1996, pp. 231–239.

[6] A. Jalali and P. Mermelstein, 'Power Control and Diversity for the Downlink of CDMA Systems', Conference records on the 2nd International Conference on Universal Personal Communications, Vol. 2, October 1993, pp. 980–984.

[7] 3rd Generation Partnership Project, 'Physical Channels and Mapping of Transport Channels onto Physical Channels (FDD)', TS. 25.211, V4.6.0, Available online at www.3gpp.org, September 2002.

[8] T. Ojanperä and R. Prasad, 'Wideband CDMA for Third Generation Mobile Communications', Artech House, 1998.

[9] A. Viterbi, 'Principles of spread spectrum communications', Addison-Wesley, 1997.

[10] F. Adachi, M. Sawahashi and K. Okawa, 'Tree-structured Generation of Orthogonal Spreading Codes with Different Lengths for Forward Link of DS-CDMA Mobile Radio', *IEE Electronic Letters*, Vol. 33, No. 1, January 1997, pp. 27–28.

[11] 3rd Generation Partnership Project, 'Spreading and Modulation (FDD)', TS. 25.213, V5.2.0, Available online at www.3gpp.org, September 2002.

[12] M. Hunukumbure, M. Beach and B. Allen, 'Downlink Orthogonality Factor in UTRA FDD Systems', *IEE Electronic Letters*, Vol. 38, No. 4, February 2002, pp. 196–197.

[13] 3rd Generation Partnership Project, 'Physical Layer Procedure', TS. 25.214, V5.4.0, Available online at www.3gpp.org, March 2003.

[14] TIA/EIA/IS-2000, 'cdma2000 Standard for Spread Spectrum Systems', 30 August 1999.

[15] TIA/EIA/IS-95-B, 'Mobile Station-Base Station Compatibility Standard for Wideband Spread Spectrum Cellular Systems', 3 February 1999.

[16] CDMA Development Group, www.cdg.org.

[17] 3GPP2 X.S0001-004-C.

[18] 3GPP2 P.S0001-B, 'Wireless IP Network Standard'.

[19] 3GPP2 P-R0001, 'Wireless IP Architecture Based on IETF Protocols'.

[20] CDMA Development Group, www.cdg.org.

[21] 3GPP2 A.S0001-A, 'Interoperability Specification (IOS) for cdma2000 Access Network Interfaces'.

[22] 3GPP TSG 29.234, V6.2.0, Feasibility Study on 3GPP System to WLAN Interworking.

[23] 3GPP TSG 22.234, V6.2.0, Requirements on 3GPP System to WLAN Interworking.

[24] 3GPP TSG 23.234, V6.2.0, 3GPP System to WLAN Interworking, System Description.

[25] IEEE Std 802.11, Wireless LAN Medium Access Control (MAC) and Physical Layer (PHY) Specifications.

[26] 3GPP2 S.R0087-0, V1.0, 3GPP2-WLAN Interworking, Stage 1, Requirements.

[27] Milind M. Buddhikot, Girish Chandranmenon, Seungjae Han, Yui-Wah Lee, Scott Miller and Luca Salgarelli, Design and Implementation of a WLAN/CDMA2000 Interworking Architecture, *IEEE Communications Magazine*, Vol. 41, No. 11, November 2003.

[28] Kalle Ahmavaara, Henry Haverinen and Roman Pichna, Interworking Architecture between 3GPP and WLAN Systems, *IEEE Communications Magazine*, Vol. 41, No. 11, November 2003.

[29] IEEE P802.15, 'IEEE P802.15 WPAN High Rate Study Group PAR', IEEE P802.15, Working Group for Wireless Personal Area Networks, February 2003.

[30] IEEE 802.11a, 'Information Technology – Telecommunications and Information Exchange between Systems – Local and Metropolitan Area Networks – Specific Requirements, Part 11: Wireless LAN Medium Access Control (MAC) and Physical Layer (PHY) Specifications, Amendment 1: High-speed Physical Layer in the 5 GHz Band', 2000.

[31] IEEE 802.16, 'IEEE Standard for Local and Metropolitan Area Networks, Part 16, Air Interface for Fixed Broadband Wireless Access Systems', April 2002.

[32] IEEE 802.16a, 'IEEE Standard for Local and Metropolitan Area Networks, Part 16: Air Interface for Fixed Broadband Wireless Access Amendment, Amendment 2: Medium Access Control Modifications and Additional Physical Layer Specifications for 2–11 GHz', April 2003.

[33] IEEE 802.20-PD-06, 'System Requirements for IEEE 802.20 Mobile Broadband Wireless Access Systems', July 2004.

[34] D. Wisely, H. Aghvami, S. L. Gwyn, T. Zahariadis, J. Manner, V. Gazi, N. Houssos and N. Alonistioti, 'Transparent IP Radio Access for Next-Generation Mobile Networks', *IEEE Wireless Communications*, August 2003, pp. 26–35.

[35] SDRF-02-A-0002-V0.00, 'Overview and Definition of Radio Software Download for RF Reconfiguration in a Technical and Regulatory Context', Software Defined Radio Forum, August 2002.

[36] J. McNair and F. Zhu, 'Vertical Handoffs in Fourth-Generation Multinetwork Environments', *IEEE Wireless Communications*, June 2004, pp. 8–15.

[37] X. Gao, G. Wu and T. Miki, 'End-to-End QoS Provisioning in Mobile Heterogeneous Networks', *IEEE Wireless Communications*, June 2004, pp. 24–34.

[38] G. Carneiro, J. Ruela and M. Ricardo, 'Cross-Layer Design in 4G Wireless Terminals', *IEEE Wireless Communications*, April 2004, pp. 7–13.

[39] ETSI TR 101 856, 'Broadband Radio Access Networks (BRAN), Functional Requirements for Fixed Wireless Access Systems Below 11 GHz: HIPERMAN', March 2003.

[40] ETSI TR 101 683, 'Broadband Radio Access Networks (BRAN); HIPERLAN Type 2, System Overview', February 2000.

[41] ETS 300 744, 'Digital Video Broadcasting (DVB); Framing Structure, Channel Coding and Modulation for Digital Terrestrial Television (DVB-T)', November 1996.

[42] R. Nee and R. Prasad, 'OFDM for Wireless Multimedia Communications', Artech House Publishers, Boston, 2000.

[43] T. Keller and L. Hanzo, 'Adaptive Multicarrier Modulation: A Convenient Framework for Time-Frequency Processing in Wireless Communications', *Proceedings of IEEE*, Vol. 88, No. 5, May 2000, pp. 611–640.

[44] M. Ergen, S. Coleri and P. Varaiya, 'QoS Aware Adaptive Resource Allocation Techniques for Fair Scheduling in OFDMA Based Broadband Wireless Access Systems', *IEEE Transactions on Broadcasting*, Vol. 40, No. 4, December 2003.

[45] T. Pollet, M. V. Bladel and M. Moeneclaey, 'BER Sensitivity of OFDM Systems to Carrier Frequency Offset and Wiener Phase Noise', *IEEE Transactions on Communications*, Vol. 43, No. 2, February 1995, pp. 191–193.

[46] J. Li and M. Kavehrad, 'Effects of Time Selective Multipath Fading on OFDM Systems for Broadband Mobile Applications', *IEEE Communications Letters*, Vol. 3, No. 12, December 1999, pp. 332–334.

[47] D. Gesbert, M. Shafi, D. Shiu, P. J. Smith and A. Naguib, 'From Theory to Practice: An Overview of MIMO Space-Time Coded Wireless Systems', *IEEE Journal on Selected Areas in Communications*, Vol. 21, No. 3, April 2003, pp. 281–302.

[48] S. M. Alamouti, 'A Simple Transmit Diversity Technique for Wireless Communications,' *IEEE Journal on Selected Areas in Communications*, Vol. 16, October 1998, pp. 1451–1458.

[49] A. J. Paulraj, D. A. Gore, R. U. Nabar and H. Bölcskei, 'An Overview of MIMO Communications – A key to Gigabit Wireless', *Proceedings of IEEE*, Vol. 92, No. 2, February 2004, pp. 198–218.

3

Data Services Architecture and Standardization

Salvador Hierrezuelo, Alejandro Gil, Juan Guerrero, Raquel Rodríguez,
Juan Torreblanca, Mattias Wahlqvist and Gerardo Gómez

3.1 Introduction

After understanding how the major cellular technologies works, the next step is to study how
services are implemented on top of them. In this chapter we will study how the overall data
services architecture is built up in cellular systems, what are the main protocols and functionalities
as well as how the different network elements interact. We will also get an introduction to
a selected set of services that altogether are a representative selection from functionality, performance, and quality of service requirements, coming from both the cellular world and the IP world.

One of the major trends of the latest years is the ramp-up of data services utilization.
Customers are more and more starting to use their wireless devices as service platforms, not
only for making phone calls, but also for sending messages, downloading mails, playing
games, etc. The increase of data service usage, the multitude of available services and the
constant increase of end-user performance requirements have started to challenge the operators
already today, and will play a more and more important role in the future. Although the data
tornado might not be here yet, it has certainly started to be windy.

3.1.1 Circuit-Switched and Packet-Switched Services

Services are traditionally split into two main categories: Circuit-Switched (CS) services, and
Packet-Switched (PS) services. The definition comes from traditional telecom vs datacom,
where telephone calls were circuits that were set up (switched) between the parties, and
computers used data packets with individual addressing to communicate between each other.
CS services are typically providing a constant bit rate and a fixed (or low variable) delay from
the moment information is sent until it arrives on the other end. Examples of CS services are

ordinary voice calls, video telephony and analog-modem connections. PS services on the other hand typically provides guaranteed delivery of the information, meaning that if there is a distortion during the transmission, the protocols will detect that and retransmit the corrupted part. This behavior leads to that PS services cannot guarantee a fixed delay in the information transfer (other than by statistical means). Examples of PS services are all services that make use of IP transport protocol.

The examples above illustrate well how the borderline between these two categories has, during the latest years, become more and more difficult to distinguish. The reason is because while we would like to classify services in terms of the added value they provide to the end-user, this classification focuses on how the information is carried through the network. This problematic is demonstrated by the development of new services like Voice over IP (VoIP), which by tradition is a CS service that is being carried by PS technology or High-Speed CS Data (HSCSD), which is a PS service which is being carried by CS technology.

As it is rather difficult to give a single and true answer on how services should be classified, we will avoid that completely in this book. Instead we will be discussing on services from a service performance perspective and nothing else. If the services should be classified later as CS or PS, if they are coming from the IP domain or from the traditional telecom domain, or if they are implemented in an intelligent network node, IMS or in a third-party server somewhere on the Internet, we will basically not worry about them as long as they do not impact on the performance of the service. For that reason, we have selected a set of key services and we will describe how they work, their performance, where they are coming from and which standards apply.

3.1.2 Services Architectures and Protocols

Mobile network operators have been providing access to services to the end-user, many of them owned by third parties (service providers). Recently, the main concern from operator's side was the lack of control over the services used by their customers. This fact has led to an evolution of the services architecture in order to develop a common infrastructure that allows network operators to support and control the services they are offering.

From an end-user performance point of view, one of the most important parts of the services architecture is the protocols that are designed to carry the data. The overhead and design of these protocols will, to a very high extent, define the limits in which the end-user perceives the performance of a given service. It is important to note that these protocols have not necessarily been designed with the target of being used over a wireless link, which certainly increases the challenge for the service provider. One example of this, which we will study in detail in this book, is the usage of TCP/IP on top of the wireless channels. Although TCP with a lot of built-in flexibility is designed to adapt to various conditions and different bandwidths, it was thought for fixed lines with fairly constant bit rate channels, low delay jitter as well as low packet-loss probability. As it will be shown later, some mechanisms in TCP may make it perform worse than expected, and be very sensitive to the characteristics of the wireless channel.

3.1.3 Services Selection

In order to study service performance, it is necessary not only to understand the general structure of cellular systems and the services architecture but also to know the details about how the protocols behave with different applications. As the amount of services is incredibly wide, with new applications coming every day, only few representative services of typical applications in wireless networks have been selected, such as SMS/MMS services, WAP and Web services, Push-to-Talk over Cellular (PoC) and network gaming.

With this selection, it is intended to capture both pure telecom services and typical services coming from the IP domain which make use of the radio network as carrier. The selection includes non-delay sensitive services such as SMS/MMS, as well as services which have very strict delay requirements like PoC and some examples of network games.

For each service it will be presented how the service functions on a general level, what protocols are used, how the signaling works and, perhaps even most important, what type of key performance indicators (KPIs) can be used to characterize the end-user performance.

3.2 Services Architecture

Traditionally in cellular networks (i.e. 2G), operators played the central and main role in the business. Cellular operator was the service provider of the one main service: CS-Voice; and at the same time controlled the carrier of the service: the cellular network. In the last years, the need or willingness to provide similar type of services available in the Internet over cellular networks drove the cellular network development.

In the old times, in fixed network, when a new service was provided the changes would affect the telecom infrastructure directly, e.g. switches. But a whole new set of services, most of them IP-based, appeared into the picture. It was not feasible anymore to modify the telecom access infrastructure for each and every new service. The philosophy followed by 2.5G and 3G tries to tackle this by decoupling the access to network from the service provisioning. The rational for this new philosophy is given by two main reasons.

1. The need to have a common access infrastructure for all the different type of services, as new services appear every day and it would not be feasible to change the access infrastructure for each new service.
2. The willingness to create new different roles in the provisioning of mobile services. New business entities, such as service providers, cellular network owners, brokers, etc., are different parties that would get revenues due to whole new set of services provided.

In this scenario, new service architecture was developed (Figure 3.1), which is compounded of cellular network operator domain (RAN/BSS and CN), service provider domain (with application servers) and the mobile terminal that accesses the cellular network and also contains application clients.

- *Cellular network owner* – is the entity that owns and manages the cellular access infrastructure, Radio Access Network (RAN) and Core Network (CN). The cellular network owner provides all the means to generate an IP access bearer that enables IP connectivity between MT and the service provider. For that purpose, the cellular network is in charge to establish

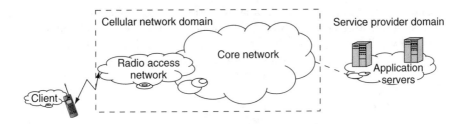

Figure 3.1 Typical service architecture (cellular network operator domain and service provider domain)

the connection over the air interface up to the CN. The bearers provided by the cellular network owner are flexible and can be customized to the different type of users or services supported (e.g. different QoS policies for each user and service user on top of the cellular network).

- *Service provider* – is the entity that owns the services contents, e.g. gaming, messaging, wireless internet, music streaming, video call, etc.
- *Mobile terminal* – is the entity that utilizes the connectivity provided by the cellular network to get access to the services provided by the service provider.

Typically, a service provider can be a mobile operator that owns the services that its users are accessing, but it can be a third-party provider different than the mobile operator. This opens a new competitive area, now that operators do not only need to compete with other operators, but also with third-party service providers. The services architecture is flexible enough to have different business cases in this regard.

In Figure 3.2, it can be observed how the cellular network domain is decoupled from the service domain. The cellular network domain provides an IP bearer service, and the user data generated by the service domain is carried on top of the IP bearer between the MT application layer and the servers in the IP external network.

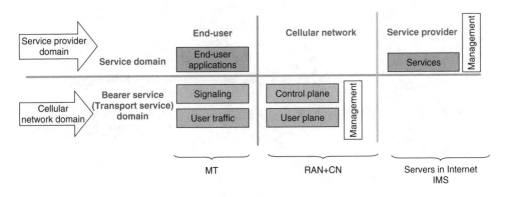

Figure 3.2 Service provider domain and cellular network domain split

The cellular network operator provides an IP bearer with QoS mechanisms. These QoS mechanisms allow to differentiate between qualities provided to a specific user and furthermore to qualities provided to different services of the same user. The flexibility provided by the IP bearer enables to adequate it to upper-level service needs.

3.2.1 Services and Service Enablers

The service bearers are established to get connectivity to certain external IP networks, in which some application servers can be encountered. The way to identify the external IP network to which the MT wants to establish the service bearer is by the Access Point Name (APN). In the UMTS world, the service bearer establishment is known as Packet Data Protocol (PDP) context activation.

In Figure 3.3, the mapping between service bearer in the cellular network domain is adapted with different QoS criteria depending on the service that is accessed (application server). There is not a specific rule of how the mapping between APNs and external IP networks shall be configured; this is let free to mobile operators' configuration.

There is another point to consider: how the data usage is charged to the clients. Currently, PS charging in the cellular network is performed based on the APNs, meaning that all IP packets of a user using a service bearer connected to a certain APN will be considered in the same way, regardless of different services and using a potentially different QoS. This is something not very desirable for mobile operators that want to have the capability to charge users differently depending on the services used by them. Currently, it is being under definition and standardization, what is called IP-flow-based charging [1], where the charging granularity is not any more APNs. This new charging system goes down to the IP flow characteristics within service bearer connected to the APN, being able to distinguish different application IP flows and charge them differently depending on operator-defined rules (e.g. streaming packets more expensive compared to Web browsing packets, etc.).

There are two main strategies for APN configuration. Those strategies are ruled by three main issues: QoS provisioning to different services, differentiated charging for different services and Mobile Station (MS) capabilities.

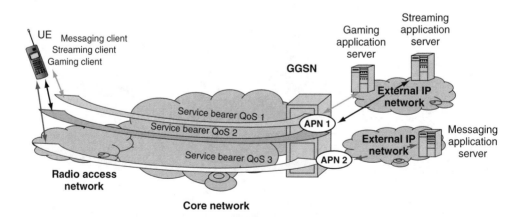

Figure 3.3 Example of mapping between service bearer and different application servers via APNs

- The first strategy consists of configuring one APN per application server. The advantage of this strategy is that it is ensured that each IP flow or service flow will be carried over different APNs and charged in a differentiated way following the mobile operator rules. The drawback is that this imposes on the MS to be able to establish several primary PDP contexts simultaneously, one per service. The reality is that currently MS terminals have a limited support of simultaneous primary PDP context, typically one, which would limit the user experience.
- The second strategy consists of configuring many application servers behind the same APN. The advantage of this strategy is that the limitation imposed by the maximum number of simultaneous primary PDP contexts in a terminal would not be affected, enabling to have access to all the potential services under the same PDP context. The drawback is that current charging systems do not allow charging granularity in the APN down to the IP flow or service flow, but the evolution of the charging systems toward the IP-flow-based charging shows that this second strategy is the future proof solution.

Normally, mobile operators use compromise solutions, where for certain services they have configured dedicated APNs, while for other services their configuration gathers together some services behind the same APN. The specific configuration will depend on the importance the operator considers the charging for specific services and the simultaneous PDP context capability of their users' terminals.

3.2.2 IP Multimedia Subsystem (IMS)

Going further in the services architecture evolution toward the full end-to-end QoS support and full end-to-end control of the connections by the mobile operator, the IMS [2] was defined.

With the split and division between cellular network domain and service domain, some of the markets that were previously belonging only to mobile operators were open to other possible business entities like service providers coming from outside the mobile operators' area. This pushed mobile operators toward the definition of the IMS, detached from the cellular network infrastructure, but defined, managed and controlled by them. This enables to expand operators' service revenues and offerings.

Apart from the business reasons, there were also other needs that triggered IMS birth. There was a clear problem from the users' perspective in the end-to-end service provisioning, mainly related to the service QoS and service control. Mobile operators could guarantee certain QoS up to their service bearer, but the external IP network is out of their control and being typically pure Internet, no QoS can be guaranteed. In order to solve this problem, IMS tackles aspects such as QoS provisioning at end-to-end level, which implies coordination between cellular network service bearer QoS and IP network QoS. QoS management and control of the different services provided by IMS enable to:

- Deliver end-to-end person-to-person real-time IP-based multimedia communication (e.g. video telephony) or person-to-machine communication (e.g. gaming)
- Enable different services and applications to interact (e.g. presence service with Push-to-Talk (PTT) service).

The architecture of IMS, depicted in Figure 3.4, is compounded of three main layers.

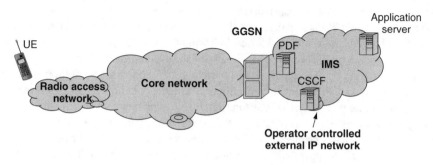

Figure 3.4 IMS architecture

1. *Connectivity layer* – formed basically by the IP backbone.
2. *Control layer* – mainly compounded by the session control entities PDF, P-CSCF, S-CSCF, etc., which are mainly responsible for controlling and authorizing the resources requested by a particular service (among other tasks).
3. *Service layer* – this is the layer containing the application servers (PoC server, instant messaging server, streaming server, etc.) the end-users are accessing to.

A more detailed description of the IMS architecture is further presented in section 4.3.3.

3.3 Data Protocols Characteristics

This section describes the main characteristics of the protocols used in wireless-data networks. The goal is to provide the reader a basic knowledge of data protocols behavior in order to understand the analysis of the service performance along the book.

Protocols described in this section belong mainly to link, network and transport layers of the Open System Interconnection (OSI) reference model (Figure 3.5), covering not only the radio protocols of the cellular technology but also the transport protocols derived from the use of Internet services. These layers are responsible for the following tasks.

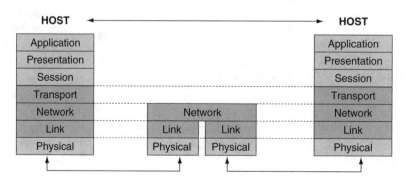

Figure 3.5 OSI protocol stack

- *Link layer* – handles the physical and logical connections to the destination and defines the format of the data. In this layer, frames are broken into bits to be sent over the network. Error detection and correction techniques are also applied at this layer.
- *Network layer* – in charge of the node-by-node addressing and routing. Routes are determined based on packet addresses and network conditions. Logical addresses are translated into physical addresses for the transmission.
- *Transport layer* – provides transparent transfer of data between peers. Main functions of transport layer are the flow and congestion control and the reliable delivery of packets.

The OSI system is a layered architecture where each layer is defined as an independent entity, which communicates with its peer through the layers settled below. Control of the information is passed from one layer to the next, starting at the application layer in one station, down to the physical layer, and then over the transmission channel to the next station and back up the hierarchy.

3.3.1 TCP/IP Networks

In the telecom world, the OSI architecture is widely accepted, but still the existence of proprietary protocols prior to the adoption of the OSI standards and their influence over settled networks makes it difficult that the whole structure can always be fully identified. However, most of the functionality in the OSI model exists currently in all communication systems, although some OSI layers may be merged into one. This is the case of TCP/IP protocol stack, developed by the US Department of Defense with the objective of connecting a number of different networks designed by different vendors into a common network. TCP/IP was adopted by the *de facto* standard for communications between different networks – Internet. Figure 3.6 shows the architecture of TCP/IP-based network.

The term 'TCP/IP' actually refers to a whole family of protocols.

- *Internet Protocol (IP)* – works at network layer and provides routing and addressing functionalities. IP runs on the end hosts as well as in the intermediate nodes with routing capabilities.
- *Transmission Control Protocol (TCP)* – provides an end-to-end pipe for the application layer to send data between peers. TCP was intended to provide a connection-oriented service with reliable transmission. TCP implements retransmission and congestion control mechanisms. Transport layer is implemented only in the hosts and not in the routers.

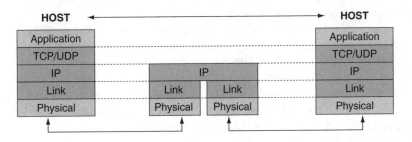

Figure 3.6 TCP/IP protocol stack

- *User Datagram Protocol (UDP)* – is an alternative transport layer protocol to TCP which also runs over IP. The UDP provides datagram service and transport addressing. However, it does not provide any reliability or congestion control. UDP is used instead of TCP over IP for certain services (e.g. streaming, WAP, etc.) which require minimum delays and may support some data losses.

Many different application protocols can use TCP/IP or UDP/IP for providing services, such as File Transfer Protocol (FTP), HyperText Transfer Protocol (HTTP) or Real-Time Streaming Protocol (RTSP).

TCP is not only very widely used by Internet applications but it is also expected to grow significantly over mobile networks due to the introduction of WAP 2.0 as well as the increasing usage of corporate office applications. This growth is expected to be more appreciable as wireless network capacity improvements demonstrate to users that they can access to all their familiar services almost as good as they can do from their fixed lines.

Being TCP/IP the technology used in the access to Internet, cellular data access had to adapt to the architecture by providing common interfaces and developing gateway functionalities between fixed and mobile domains. However, TCP/IP protocol mechanisms do not always fit very well to the inherent characteristics of wireless paths, and the study of these interactions have become a key area of research due to its important impact on the services performance.

The following sections describe the main characteristics of TCP/IP protocols, flow control, congestion control, error recovery, etc., whereas the interactions and implications of these characteristics with user performance will be further studied in following chapters.

3.3.1.1 TCP Characteristics

TCP is a transport protocol designed to provide reliable transfer of data over an unreliable datagram network. IP provides data-stream transmission and routing capabilities, but does not guarantee that data is delivered correctly. TCP uses retransmissions based on acknowledgments (ACKs) and timers in order to provide data reliability. Additionally, TCP must ensure that there is flexibility in the communication between different users, provide the capabilities to support multiple parallel connections as well as control the flow of data and congestion in the network.

Several TCP functions can be identified in order to provide the desired behavior:

- *Connection-oriented protocol*: Before starting the transmission of data, a negotiation of connection parameters has to be done between the two remote peers.
- *Segments and sequence numbers*: Data is segmented in chunks called *segments*, which are identified by a sequence number. Those sequence numbers allow identifying the information, reconstructing the initial data and detecting the need for retransmission.
- *Multiplexing*: To allow for many processes within a single host to use TCP communication simultaneously, the TCP provides a set of addresses or ports within each host. A TCP connection is fully identified by two pairs of IP address and port numbers (sockets), which correspond to the server and client side of the connection.
- *Retransmissions*: In case data is not received for some time or data loss is identified, TCP will retransmit the data segments till they are all acknowledged.

- *Flow control*: In order to minimize the impact of data losses and retransmissions, the amount of data under transmission is controlled by means of adaptive transmission window.
- *Congestion control*: Congested networks will discard packets by increasing the probability of retransmissions and degrading the end-user performance. Each TCP connection will try to adapt the size of its individual transmission window to minimize the probability that there is buffer overflow in any of the routers in the communication path [49].

Table 3.1 contains the definition of some basic concepts generally used in TCP.

Connection-Oriented Protocol

TCP is a connection-oriented protocol, which implies that a peer-to-peer connection must be established before any application data is sent. Once the transmission is finished, the connection is released.

The signaling flows of the TCP connection establishment and release scenarios are depicted in Figure 3.7.

TCP connection establishment follows the well-known 'three-way handshake' procedure, which takes approximately one-and-a-half RTTs (Figure 3.7). The initial sequence number that each terminal selects for the transmission is communicated to the other end in the respective SYN segment during the connection establishment. Additionally, during the set-up phase both ends agree on the size of the largest segment that can be used within that connection,

Table 3.1 TCP concepts definition

Name	Description
Segment	TCP data block.
MSS	Maximum Segment Size allowed by the receiver. Default value is 536 bytes, and it does not include the TCP/IP headers.
Congestion window (cwnd)	Variable that limits the amount of data that a TCP connection can send. Congestion control algorithms will change this value during the TCP connection. It is limited by the advertised window size.
Initial Window (IW)	Size of the sender's congestion window after the three-way handshake is completed.
Advertised window	Maximum value of the receiver window. This value is indicated by the client during connection establishment and limits maximum amount of unacknowledged data the sender is allowed to have under transmission during the whole connection. It is used as upper limit of the congestion window.
Slow start threshold (ssthresh)	If $cwnd \leq ssthresh$, slow start algorithm is in use. Otherwise congestion avoidance is considered. In the beginning of a connection $ssthresh = advertised window$.
Flight size	It is the amount of data that has been sent but not yet acknowledged. Bytes that are acknowledged by SACK option (if present) do not decrease flight size till every byte between them and the beginning of the transmission window are acknowledged.
Round-Trip Time (RTT)	Estimated time spent from TCP segment transmission to TCP segment ACK.
Retransmission Timeout (RTO)	Time TCP layer waits before retransmitting a TCP segment.

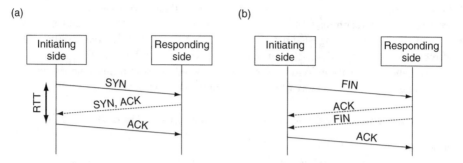

Figure 3.7   TCP connection establishment (a) and release (b)

Table 3.2   Default MTU size for different media

Network	MTU (bytes)
16 Mbps token ring	17914
4 Mbps token ring	4464
FDDI	4352
Ethernet	1500
IEEE 802.3/802.2	1492
X.25	576

known as Maximum Segment Size (MSS). Maximum value of the MSS is normally limited by the Maximum Transfer Unit (MTU) defined at link layer.

$$MSS = MTU - TCP\_header - IP\_header$$

As an example, considering a typical MTU value of 1500 bytes (Ethernet network) and 40 bytes of TCP/IPv4 headers, MSS value would be 1460 bytes. The reader can find a list of Default MTU size for different media in Table 3.2. Once there is no more data to transmit, both terminals must send a FIN segment followed by the corresponding acknowledged by the other end in order to close the connection.

From the performance point of view, establishment delays and overhead of data segments are two important factors to consider. In wireless environments with high latencies, the time needed for the establishment of a TCP connection can affect very much the performance perceived by the users, especially when the profile of the traffic is bursty or when initial delay is a key factor. Also due to bandwidth limitation of the systems, keeping the header overhead to minimum percentage (i.e. using high packet sizes) is normally a good strategy to improve the transfer rates. Of course, in systems with very unreliable transmission in the lower layers, having smaller segments can help to minimize the retransmission overhead, thus a trade-off solution shall be found.

Segments and Sequence Numbers
Once TCP connection is established, TCP divides data into segments before being transmitted to lower layers, according to the negotiated MSS.

TCP assigns a sequence number to every byte. The sequence number of the first byte in the segment is transmitted in the header of the segment. The receptor must acknowledge the received segment by sending back the sequence number of the last byte received in the segment plus 1, i.e. the next sequence number the receiver is waiting for.

In the client side, the segments are received and data is acknowledged by sending ACK segments which contain the sequence number of the last data received in sequence. ACK information is typically sent after reception of every second segment or after a timer expires (i.e. delayed TCP ACK). If the communication is full duplex and there is data to transmit in the opposite direction, the ACK information is included in the header of data segments, otherwise ACK segments including only TCP header information are sent. Typically this ACK segments are used also by the server to estimate the latency of the network, and adapt the TCP transmission rate and retransmission timer to the bandwidth available in the network.

It is also important to note from the performance point of view that the ACK information generated by TCP connections tends to be composed of lots of small packets (typically 40 bytes) which generate a regular bursty traffic pattern, which can cause delays and overheads derived from radio resource allocations in wireless systems.

Multiplexing

Not only different users but also services of the same users can have their own TCP connections, which will have to share the available resources in the lower layers. TCP provides the necessary identification capabilities through the sockets (IP address, port number) to fully identify each connection.

However, in cases where the bandwidth available is limited, having several parallel connections will also result in a degradation of the overall service. The establishment overheads and flow control will contribute to limit the performance of each individual connection. For an optimal utilization of the available bandwidth, especially critical in wireless environments, application services should try to minimize the number of parallel TCP connections that they will manage. However, from optimization point of view, it must be noted that system with higher bandwidths can still get benefits from parallel TCP connections.

Retransmissions

Standard TCP implementations follow a cumulative approach in the ACK process. This is the default behavior of TCP when no advanced features are used. With the cumulative approach, the ACK of a segment implies the ACK of all the previous segments, i.e. the ACK of a sequence number implies the ACK of all the bytes with smaller sequence number. If a segment loss occurs, every new segment received out of sequence will trigger the transmission of a new ACK segment, which will keep informing only about the information received in sequence. These messages are known as 'duplicated' ACK and are used as an indicator by the transmitter to detect problems in the transmission of the data. The transmitter is then aware of segment losses in the network when several duplicate ACKs arrive from the receiver (i.e. fast retransmission/fast recovery mechanism [3]).

Transmitter also uses a retransmission timer to control the loss of communication with the other end. This timer is reset every time a new ACK arrives to the transmitter. The maximum value of the retransmission timer is calculated dynamically based on estimations of the RTT experienced by the segments and associated ACKs. When the retransmission timer expires, all the information in the transmission window, which is not yet acknowledged is retransmitted. Retransmission timer expiration value is known as RTO.

Flow Control

TCP controls the amount of information injected in the network in order to avoid incrementing too much the congestion and to minimize the impact of retransmissions. If Retransmission Timer expires, all the information transmitted but not yet acknowledged will be retransmitted, even if part of it was already received. However, too small transmission windows will limit the connection throughput even if the network would allow higher bit rates.

The receiver side includes in every ACK segment a field with the 'advertised window' for the transmitter, indicating the maximum amount of data that it is able to receive. This value sets the higher limit for the transmitter's window. Normally, the maximum size of the advertised window is a parameter which can be defined within the operation system. If the advertised window is filled, the transmitter can send more segments only after it receives ACKs from the receiver. The advertised window is often also referred to as Receiving Window (RWIN).

Congestion Control

The objective of congestion control is to avoid the sender to introduce too much data into the network, which could cause buffers in some bottleneck router to overflow and start dropping packets.

TCP adapts its transmission bit rate based on the probability that TCP segments are discarded by the network when the load is high. The design was done based on the assumption that transmission errors occur rarely and therefore a packet loss implies congestion. This assumption is very likely for fixed networks, but when working with wireless networks, the probability that data is lost in the lower layers grows. Not only radio loss affects, which can be highly improved with acknowledged modes and retransmission schemes in the Radio Link Control (RLC) layers, but also because of errors or discarded frames in the core of the radio network.

Transmission rate of a TCP connection is limited by means of the congestion window, which defines the maximum amount of unacknowledged data that the transmitter can send. The size of this window is controlled by both transmitter and receiver in two different ways. The maximum value of the congestion window is set by the receiver end by means of the advertised-window parameter, which is sent during the establishment of the connection, and in every TCP ACK segment. On the other hand, the transmitter side increases, decreases or maintains the size of the window based on the acknowledged information received from the TCP client. At any time, advertised window from the receiver side will limit the maximum amount of data under transmission. Therefore, minimum of both windows limits the amount of outstanding data, i.e. the amount of data which is somewhere in between source and destination, and it has not been acknowledged.

Classic TCP implementations [4] include several congestion control mechanisms, being the most relevant ones the slow start, congestion avoidance and fast retransmission/fast recovery.

Slow Start

The purpose of slow start is probing the network for possible congestion and controlling the speed at which the sender introduces data. The congestion window in the sender side starts with a certain minimum value, typically 1–4 times the MSS, and it is successively increased to allow higher TCP throughput. Every time a new ACK arrives, the congestion window slides so that all the acknowledged data gets out of the window. Also, the size of the window is increased by 1 MSS. Immediately, a number of segments can be sent, such that the amount of data under transmission is equal to the congestion window size.

The basis of the sliding and growth process can be summarized as follows (Figure 3.8).

- Sender transmits a new segment.
- Sender waits for the ACK from the receiver.
- If the ACK is received before a timeout occurs, slide the window forward and increase the window size (if advertised window is not exceeded).
- If there is a timeout, retransmit the unacknowledged segments and decrease the window size.

The way the transmitter increases and decreases its window depends on the congestion control algorithm driving the transmission. At the beginning of the transmission, the network status is unknown. In this situation TCP probes the network slowly in order to determine the available capacity and in order to avoid producing congestion in network. The congestion control algorithm in charge of this initial phase of the transmission is known as slow start. Slow start is used at the beginning of a transfer or after an RTO in order to test the network.

The Initial Congestion Window (IW) is the size of the window, in terms of segments, that the transmitter uses at the beginning of the slow start phase. First TCP flavors defined one segment as IW although currently most implementations tend to use two segments. According to the initial IETF recommendations on TCP congestion control [4], IW must be less than or equal to $2 \cdot$ MSS bytes and must not be more than two segments. However, further extensions of the standard allow a larger IW [5]. With this extension, the transmitter might use 3 or 4 segments as IW (without exceeding 4380 bytes).

$$IW = \min(4 \cdot MSS, \max(2 \cdot MSS, 4380 \text{ bytes}))$$

Slow start continues to increase the size of the congestion window as long as the maximum size of the window, settled by the advertised window, is not reached, slow start threshold is not exceeded, or there is no retransmission event. If the congestion window grows reaching the slow start threshold then the congestion avoidance algorithm is used for the transmission.

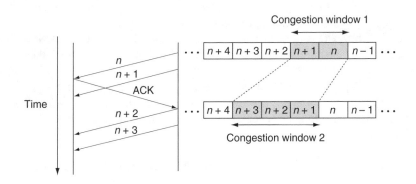

Figure 3.8   TCP sliding window

Therefore, the rule to determine which algorithm governs the transmission is based on the congestion window and the slow-start threshold. During congestion avoidance, the increase of the congestion window is linear, i.e. congestion window is incremented by 1 MSS segment per RTT.

The growing time needed by slow start to reach its maximum capacity will depend very much on the frequency at which ACKs are received (basically the RTT). For wireless networks this can be an important limiting factor for the throughput. It is proposed for example to use higher initial values for the congestion window (2, 3 or 4 MSS) in order to speed up the slow start period. But in any case this improvement has to be considered quite carefully because this speed up of the congestion window can also be quite dangerous if there is already some congestion in the network. Normally, a maximum of 2 MSS is recommended [5].

Congestion Avoidance
Congestion avoidance is controlled by the slow start threshold that initially is set to the maximum value allowed by the advertised window. During slow start, the congestion window size increases exponentially till congestion is reached and some packet is dropped. After this happens, the slow start threshold and congestion window are decreased and the congestion avoidance will start. During this procedure the congestion window is incremented more slowly (as shown in the following equation) till a maximum capacity of the network is achieved.

$$cwnd = \left(cwnd + \frac{MSS \cdot MSS}{cwnd} \right)$$

This procedure continues until congestion window reaches the limit given by the advertised window or until a segment needs to be retransmitted.

Retransmission Timeouts
The basis for the TCP congestion detection is the retransmission timer. After a retransmission due to timer expiration (RTO), TCP considers congestion in the network and tries to alleviate it. In that situation, congestion window is reduced to one segment, which decreases drastically the transmission rate, in order to immediately reduce the congestion. Slow start threshold is also decreased to half of the outstanding data. After this reset of the congestion window, as congestion window is smaller than the slow start threshold, slow start is the algorithm used after the timer expiration (the whole procedure is shown in Figure 3.9).

The value of the RTO is not constant, but it is dynamically computed based on the RTT values measured for the different segments. Since RTT is also changing along the time due to different network effects, RTO will try to adapt dynamically to the status of the network, reacting quickly to potential packet losses but also avoiding unnecessary retransmissions.

First TCP specification includes Jacobson algorithm to compute RTO. This algorithm is a simple low-pass filter of the instantaneous RTT along the time. Exact computation is based on the following algorithm [6]:

Step 1 Two state variables are maintained: Smoothed Round-Trip Time (SRTT) and round-trip time variation (RTTVAR), and a clock granularity of G seconds are assumed.
Step 2 *Initial RTO* = 3 s

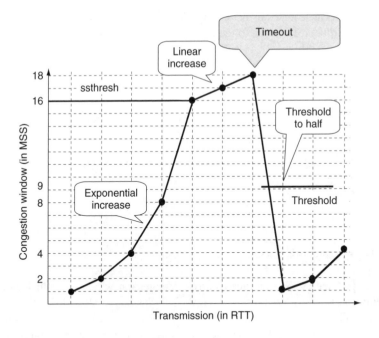

Figure 3.9 TCP Retransmission Timeout [35]

Step 3 When the first RTT measurement is made:

$$SRTT = RTT$$

$$RTTVAR = RTT/2$$

Step 4 When a subsequent RTT measurement is made:

$$RTTVAR = (1 - \beta)^* RTTVAR + \beta^* | RTTVAR - RTT |$$

$$SRTT = (1 - \alpha)^* SRTT + \alpha^* RTT$$

The above 'should' be computed using $\alpha = 1/8$ and $\beta = 1/4$

$$RTO = SRTT + \max(G, 4^* RTTVAR)$$

Step 5 When a timeout occurs, TCP doubles the value of the RTO since TCP considers that there is congestion in the network, up to a maximum of approximately 1 minute, depending on TCP implementation. Minimum value of RTO is also limited to 1 s.

Fast Retransmit/Fast Recovery

This algorithm is aimed to recover from isolated packet losses without having to wait for RTOs and reducing the penalty that timeouts have over slow start. The idea behind this algorithm is the detection of duplicated ACKs (frequently called DUPACKs). Due to cumulative ACK scheme a duplicated ACK is produced when there is a gap in the segment reception sequence and an out of order segment triggers a new ACK.

Duplicated ACKs are counted in the sender's side, and if *three* consecutive events are detected, the fast retransmit/fast recovery is triggered, performing the following several actions.

Step 1 First unacknowledged segment is retransmitted.
Step 2 *ssthresh* is readjusted according to a new value according to the formula:

$$\text{ssthresh} = \max\left(\frac{1}{2} \cdot \text{flight size}, 2 \cdot \text{MSS}\right)$$

Step 3 *cwnd* is also reduced to a lower value, but higher than the initial, to reduce the TCP transmission rate.

$$\text{cwnd} = \text{ssthresh} + 3 \cdot \text{MSS}$$

Step 4 Retransmission timer is re-started.

If more duplicated ACKs continue to arrive, congestion window is still incremented by MSS, within the limits of advertised window, in order not to slow down the transmission rate. When finally an ACK arrives for new data, the fast recovery algorithm sets the cwnd to the value of ssthresh in order to avoid the slow start phase, and transmission is continued in congestion avoidance state. If retransmission timer expires, then the first unacknowledged segment is transmitted again and default retransmission algorithm due to timeout is followed by the sender. The whole procedure is shown in Figure 3.10.

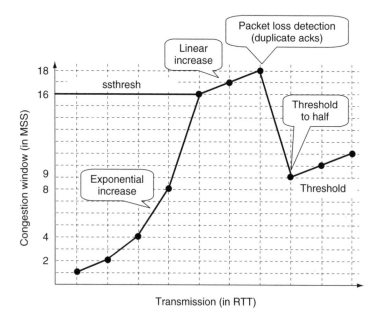

Figure 3.10   TCP packet loss detection [35]

UDP Characteristics

UDP provides an unreliable connectionless delivery of data over a datagram network. UDP does not perform any of the features described previously for TCP: segmentation, retransmissions, ACKs, flow control and congestion control. Therefore, as the underlying IP layer, UDP packets can be lost, duplicated, arrive out of order or discarded if they arrive faster than the destination process capacity. With these characteristics, UDP might use all the available bandwidth at anytime, which would be good for certain services.

However, UDP presents some important drawbacks, such as the probability of packet losses, which makes UDP to trust in the reliability of the lower layers, or the absence of congestion awareness, which could contribute without control to network congestion collapse. Additionally, the absence of flow control could overflow the receiver buffers if the data rates are high.

From cellular point of view, UDP is much less problematic than TCP in terms of performance. First, UDP protocol overhead is significantly lower than TCP (UDP header size is 8 bytes compare to TCP header of 20 bytes). Secondly, UDP does not require ACKs from receiver to transmitter and does not use any congestion or flow control mechanism.

Therefore, UDP is better protocol for transporting real-time traffic like multimedia streaming (e.g. video and/or audio) or conversational services (e.g. VoIP and videoconference). In general, this type of services requires low delays although they tolerate some packet losses. In case of packet losses, the best option from the application point of view is to discard the packets instead of recovering them by means of retransmissions, which would lead to degradation in the interactivity of the service. Therefore, packet losses may be directly mapped into the quality of the received media.

3.3.2 Impact of Radio Interface on Transport Protocols

Characteristics of the link layer of wireless networks (i.e. 2.5G/3G) have significant effects on TCP and application layers performance. Big latencies, variable data rates, asymmetry, delay spikes, data loss and bandwidth oscillations are some of the main effects that can affect to the performance of TCP [34]. For this reason, it is important to understand how these different effects interact with the protocol, in order to find out possible improvements or parameter optimizations that could help to reduce any degradation. A summary of those effects is described below (further details can be found in section 5.4.1.1).

- *Latency*: The latency of wireless links is normally high due to the extensive processing required at the physical layer of those networks.
- *Data Rates*: Limited data rates and bandwidth oscillation due to effects from other users and from mobility.
- *Asymmetry*: 2.5G/3G systems typically have asymmetric uplink and downlink data rates.
- *Delay spikes*: 2.5G/3G links are likely to experience sudden variations in the delay, exceeding the typical RTT by several times due to different reasons (section 5.4.1.1).
- *Packet loss due to corruption*: Wireless systems present a low level of packet losses thanks to the link level retransmissions performed, but still the transmission errors will be seen as a delay jitter by the upper layers.

3.4 SMS/MMS

3.4.1 Introduction to SMS

The purpose of the Short Message Service (SMS) is to provide the means to transfer text messages from a mobile device to another mobile device or a device connected to a fixed network via an SMS center (SC or SMSC). This way, SMS are not sent directly from sender to recipient, but always via the SC, which means that each mobile telephone network that supports SMS has one or more messaging centers to handle and manage the short messages.

These messages may be up to 160 characters of text in length which can be comprised of words, numbers or an alphanumeric combination. The final number of characters depends on the data coding schemes used for the text part of the message. The three text coding schemes that can be used in SMS are the GSM 7-bit default alphabet [7], 8-bit data, and the Universal Character Set (UCS2) [8], which uses 16 bits per character. Since the text messages contain up to 140 octets, 160, 140 or 70 characters may be sent in one message when GSM 7-bit, 8-bit data or UCS2 are used respectively. The GSM 7-bit default alphabet is composed of 128 characters plus 9 additional characters (extension table) including the Euro sign. The UCS2 with 2-byte symbols is used for encoding complex sets of non-Latin characters such as Chinese and Arabic.

SMS can be sent and received via GSM or GPRS bearers. In case GSM is used as bearer, short messages travel over the radio channel using the signaling path. As such, users of SMS rarely ever get a busy signal as they can do during peak network usage times. However, when GPRS is used as bearer, the SMS is supported by a packet data traffic channel. An active mobile device is able to send or receive a short message at any time, independently of whether or not there is a speech or data call in progress. The usage of the SMS bearer (GSM or GPRS) is determined by the operator network capability and settings. Users are usually given the choice to select GPRS to be the bearer. When GPRS is selected as SMS bearer, the MS tries first to send the SMS over GPRS. If the sending fails, the MS sends it over GSM.

SMS also features confirmation of message delivery, which means that users can receive a return message back notifying them whether the SMS was delivered or not. When a mobile device submits a short message, a report is always returned to the mobile device either confirming that the SC has received the short message or informing that it was impossible to deliver the short message to the SC, including the reason. This procedure is also performed when it is the mobile device that receives a short message from the SC. The SMS also offers to the SC the capabilities of informing the MS of the status of a previously sent mobile-originated short message:

- Successfully delivered to the recipient.
- The SC was not able to forward the message to the recipient. The reason can be an error of permanent or temporary nature.

  - Permanent errors can be, for example, validity period expired, invalid recipient address.
  - Errors of temporary nature can be, for example, SC-recipient connection being down, recipient temporarily unavailable.

The short message comprises seven elements related to the submission and reception of messages, which are described in Table 3.3.

Table 3.3   Short message information elements

Element	Description
Validity period	Allows the mobile device to indicate a specific time-period value in the short message for which the short message is valid, i.e. for how long the SC shall guarantee its existence in the SC memory before delivery to the recipient has been carried out.
Service center timestamp	By this information element, the SC informs the recipient about the time of arrival of the short message at this SC.
Protocol identifier	Indicates which is the application layer protocol used or interworking with a certain type of telematic device. For example, an SMS message which is required to interwork with Internet email may have its protocol identifier value set for Internet email.
More messages to send	Used by the SC to inform the mobile device that there are one or more messages waiting in the SC to be delivered to it.
Delivery of priority and non-priority messages	Indicates the network whether or not a message is a priority message. When a short message is a non-priority message, its delivery is not attempted if the mobile device has been identified as temporarily absent, although it is attempted if the MT is identified as having no free memory capability. On the other hand, the delivery of a priority message is always attempted irrespective of whether or not the MS has been identified as temporarily absent, or having no free memory.
Messages-waiting	Enables the network to provide the HLR, SGSN and VLR (with which the recipient mobile device is associated) information about a waiting message in the originating SC. This is only used in case of previous unsuccessful delivery attempt(s) due to temporarily absent mobile or memory capability exceeded in the mobile device. The HLR may contain a list of addresses of SCs which have made previous unsuccessful delivery attempts of a message, so that it can send alert messages to them indicating that a mobile device has recovered operation (e.g. has responded to a paging request) or has memory capacity available to receive one or more messages. In case the HLR does not contain that list and the delivery of a short message could not be carried out successfully, the SC is notified of the reason of the mobile device being absent and that SC retransmits the short message periodically in order to ensure delivery.
Alert-SC	Used to inform the SC that an MS to which a delivery attempt failed, is now prepared to receive the message because has resumed operation or has memory newly available.

3.4.2 SMS Architecture and Signaling

The basic network architecture of the SMS is depicted in Figure 3.11 [9].

The SC is responsible for relaying, storing and forwarding of a short message between the entities sending and receiving the SMS (e.g. MS). One SC may be connected to several PLMNs, which may be connected to several MSCs (SMS-GMSCs or SMS-IWMSCs) within one and the same PLMN. The SC may be also be integrated with the SMS-GMSC/ SMS-IWMSC.

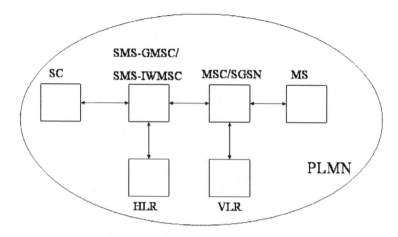

Figure 3.11   Network architecture of the SMS

The SMS-GMSC is a function of an MSC which is in charge of the reception of the short message from the SC, interrogation to the HLR to obtain routing information and transferring the short message to the MSC or SGSN of the recipient MS.

The SMS-IWMSC is a function of an MSC which receives the short messages from the MSC or SGSN, establishes, where necessary, a link with the addressed SC and transfers the short message to that SC. If a delivery confirmation associated with the short message is received from the SC, the SMS-IWMSC is responsible for the relaying of the confirmation to the MSC or SGSN.

Two sequence diagrams for a successful short message retrieval attempt, for both mobile-terminated and mobile-originated SMS, are shown in Figure 3.12.

For the mobile-terminated short messages, the SMS-GMSC receives a short message from the SC. If the SMS-GMSC finds any error within the parameters of the message, it returns the appropriate error information to the SC in a failure report; otherwise it interrogates the HLR to retrieve routing or possible error information. When the HLR returns error information (e.g. the mobile device is absent) the SC is notified with a failure report, otherwise the short message is sent to the MSC (or the SGSN) of the recipient's mobile device. When the HLR is interrogated, it can return one or two addresses (MSC, SGSN or both). If the SMS-GMSC receives two addresses, it can try using the second path once the first one did not succeed. When an MSC receives the short message that MSC retrieves information from the VLR (e.g. location area) and if no errors are indicated from that VLR, the message is transferred to the MS. On the other hand, if an SGSN receives the short message, it sends the short message to the MS directly. Both elements, VLR and SGSN, can detect errors in the short message transference procedure. If that happens, the appropriate error information is returned to the SMS-GMSC in a failure report, and the transference is aborted. When receiving a confirmation that the message is received by the MS, these elements send the delivery confirmation to the SMS-GMSC in a delivery report, else if a failure report is received, the appropriate error information is sent to the SMS-GMSC in a failure report.

Finally, the SMS-GMSC sends a delivery or a failure report to the SC depending on the success of the transference of the short message and also indicates the HLR the result of the transference(s) and the reason of the failures, whenever necessary (for the first, second or both attempts).

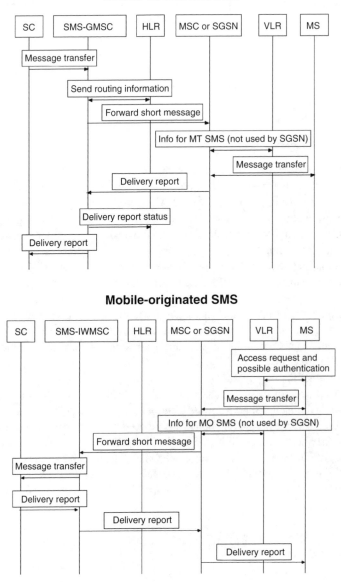

Figure 3.12   Example of successful short message transfer attempt

When an MSC receives a notification from a MS that it has memory available to receive one or more short messages, it relays the notification to the VLR. If errors are indicated by that VLR, the appropriate error indication is sent to the MS in a failure report. On the other hand, if it is the SGSN which receives this notification, it relays the notification to the HLR, in the same way that it does when it receives a notification that the mobile station is becoming reachable again.

For the mobile-originated short messages, either the MSC or the SGSN receive a short message from the MS. If the MSC receives that message, it first retrieves information from the VLR (e.g. the MSISDN for the MS) so that if errors are indicated by that VLR, a failure report is returned to the MS. If no parameter errors are found by the MSC or SGSN, the short message is sent to the SMS-IWMSC. The SMS-IWMSC establishes, where necessary, a link with the addressed SC and transfers the short message to that SC. If a report associated with the short message is received from the SC, the SMS-IWMSC is responsible for relaying the report to the MSC or SGSN. If no report is received before a timer expires or if the SC address is invalid, the SMS-IWMSC returns the appropriate error information to the MSC or SGSN in a failure report.

3.4.3 SMS Protocol Stack

The SMS protocol stack consists of four layers: the application layer, the transfer layer, the relay layer and the link layer [9]. These protocol layers are structured as shown in Figure 3.13.

The Short Message Application Layer (SM-AL) is implemented in devices in the form of software applications that send, receive and interpret the content of messages (e.g. message editor, games, etc.). This layer would also perform tasks as the concatenation of SMSs, which is a feature that is not usually available in all the mobile devices.

Any application that is created to send, receive or just handle SMS shall be built on top of the Short Message Transfer Layer (SM-TL), which provides a set of primitives to the SM-AL. This service enables the SM-AL to transfer short messages to other devices (e.g. mobile phones or computers), receive short messages from them and receive reports about earlier requests for short messages to be transferred. At this level, messages are considered as organized sequence of bytes containing information such as message length, originator or recipient, etc. All transactions are acknowledged between the mobile device and the network in order to cope with message losses.

In order to keep track of messages and reports about those messages, primitives between the SM-AL and SM-TL contain a Short Message Identifier (SMI), which is a reference number for the message associated with the primitive. This SMI is mapped to and from the identifier used between the SM-TL and the Short Message Relay Layer (SM-RL). The SMI is not carried between entities and therefore a given message may have different SMIs at the MS and SC sides [9].

The SM-RL allows the transport of messages and reports between various network elements. Both the MSC and the SMS-GMSC/SMS-IWMSC work at this level. A network

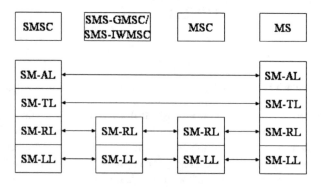

Figure 3.13   SMS protocol stack

element may store temporarily a message if the next element to which the message is to be forwarded is not available. Like primitives between the SM-AL and SM-TL, primitives between the SM-TL and SM-RL contain an SMI, which is a reference number for the data units associated with the primitive that allows the tracking of messages and reports about those messages.

The Short Message Link Layer (SM-LL) allows the transmission of the message at the physical level. For this purpose, the message is protected for coping with low-level channel errors.

3.4.4 Introduction to Multimedia Messaging Service (MMS)

After the SMS success in the GSM 2G system, the support of higher data rates and the introduction of the packet switch technology in 2.5G and 3G networks have allowed the development of alternative and advanced data services. The increasing user demand for multimedia features, together with the need for a convergence with the existing services available in the Internet have led to the development of new messaging services.

MMS is the natural evolution of SMS and allows users and machines to send and receive messages exploiting the whole array of media types available today, e.g. text, photos, images, audio and video. All these media types can be sent together within the same message, while also making it possible to support new content types as they become popular.

Apart from being a major driver for person-to-person communication, MMS is important as a booster for other 3G services. For the operators, MMS is a convenient intermediate step from SMS on the way to launch other new services like video streaming.

MMS has been specified to interact with other messaging systems. It is possible to send MMs (Multimedia Messages) to other MMS-enabled phones, to email addresses and even to facsimile addresses. Users of legacy terminals without MMS capability are also able to access the service, as these users can receive in their terminals an MMS notification via SMS of a specific Internet address where the message can be retrieved.

The MMS standardization process has been mainly carried out by the 3GPP and the WAP Forum [10]. The 3GPP is responsible for the high-level service requirements, architectural aspects of MMS, message structure and content formats, whereas the WAP Forum deals with the technical realizations of the interface, bridging the mobile device and the network on the basis of WAP and Internet transport protocols. The first technical realization is based on WAP transport protocols and is named MMS 1.0. The second technical realization is based on either WAP or Internet transport protocols and is named MMS 1.1.

3.4.5 MMS Architecture and Signaling

Multimedia messaging may encompass many different network types. The basis of connectivity between these different networks is provided by the Internet protocol and its associated set of messaging protocols. This approach enables messaging in 2G and 3G wireless networks to be compatible with messaging systems found on the Internet. Figure 3.14 shows the main MMS architectural elements [11].

The MMS network architecture encompasses all the various elements which are needed to deliver a complete MMS to a user, including interworking between service providers. It also combines and integrates messaging systems that were already existing in different networks.

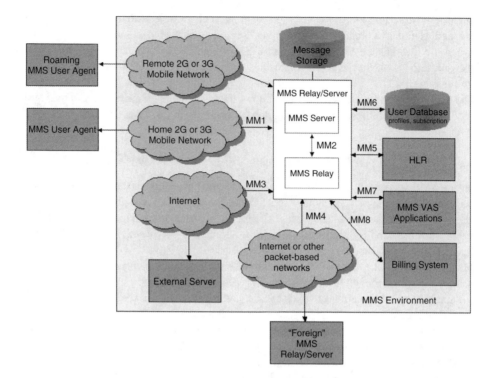

Figure 3.14   MMS Architectural Elements

The MMS Environment (MMSE) is a set of MMS-specific network elements under the control of a single administration (MMS provider), which is responsible for providing the service (e.g. submission, delivery, storage and notification functionality) to the MMS subscribers. The MMSE may comprise 2G and 3G networks, 3G networks with islands of coverage within a 2G networks and roamed networks. In the case of roaming, the visited network is considered a part of that user's MMSE. The MMS-enabled terminals operate with the MMSE to access the service. Subscribers to another service provider are considered to be a part of a separate MMSE.

The MMS User Agent resides on a User Equipment (UE) or on an external device connected to a UE. It is an application layer function that provides the users with the ability to view, compose and handle MMs (e.g. submission, reception and deletion of MMs).

Like in the SMS system, MMs are not sent directly from the sender to the recipient. The exchange of MMs is always done through the MMS Relay/Server. Some of its most important functionalities are the reception and sending of MM, the conversion of messages from legacy messaging systems to MM format and vice versa, the retrieval of messages, the sending of notifications to the MMS User Agents, the routing of forward MMs and generation of read-reply reports, address translation, transfer of messages between different messaging systems, etc.

Depending on the business model, the MMS Relay/Server may be a single logical element, which is then known as MMS Center (MMSC), or may be separated into MMS Relay and MMS Server elements. In the later configuration, the MMS Relay is responsible for routing-related tasks within and outside the MMSE, whereas the MMS Server is in charge of storing messages that are waiting for retrieval. The MMS Relay/Server is able to generate charging data (Charging Data Record, CDR).

The MMS User Database may be comprised of one or more entities that contain user-related information such as subscription and configuration (e.g. user profile). This MMS User Database and the HLR hold information for the access control to the MMS, control of the extent of available service capability (e.g. server storage space), a set of rules how to handle incoming messages and their delivery and information of the current capabilities of the terminal.

The Value Added Services to MMS (MMS VAS) applications offer additional services to MMS users, like presence service, weather updates, etc. There could be several MMS VAS applications included or connected to an MMSE. MMS VAS applications may also be able to generate CDRs.

MMS specifications define a set of reference points in the MMS architecture, which are described in Table 3.4 and shown in Figure 3.14.

Table 3.4 Reference points in the MMS network architecture

Reference point	Description
MM1	Reference point between the MMS User Agent and the MMS Relay/Server. This reference point is used to submit and retrieve MMs, to send notifications and to exchange delivery and read-reply reports between MMS Relay/Server and MMS User Agents. So far, the implementation of MM1 has been based on WAP, but other implementations based on applications using MExE [12] and other standardized Internet protocols (like Session Initiation Protocol, SIP) will be carried out in the future.
MM2	Reference point between the MMS Relay and the MMS Server. Currently there is no technical realization for this interface in the MMS standards and most of the commercial solutions offer a combined MMS Relay and MMS Server in the form of an MMSC. The implementation of MM2 is proprietary.
MM3	Reference point between the MMS Relay/Server and the external servers that allows the sending and retrieval of MMs from servers of external messaging systems, e.g. email servers and SMSCs. The content and structure of messages can be adapted in order to guarantee interoperability with legacy messaging systems. This adaptation can be performed by the MMSC or another network entity.
MM4	The MM4 interface is an interface between two MMS Relay/Servers belonging to different MMSEs that allows the transfer of messages between them. The implementation of this interface is based on SMTP.
MM5	Reference point between the MMS Relay/Server and the HLR that allows the MMS Relay/Server to interrogate the HLR for obtaining information about a particular subscriber and routing information for forwarding a message to another messaging domain. In case of using SMS as the bearer for notification, this reference point is not necessary.
MM6	Allows interactions between the MMS Relay/Server and user databases (e.g. users' profiles, subscription parameters). This interface has not been standardized yet.
MM7	Reference point between the MMS Relay/Server and the MMS VAS applications, which allows the interchange of MMs between those network elements. This interface is based on SOAP 1.1 and SOAP messages with attachments using an HTTP transport layer, although other implementations based on different transport layers could be used in the future.
MM8	Allows interactions between the MMS Relay/Server and the billing system. Like the MM6 interface, this interface has not been standardized yet.

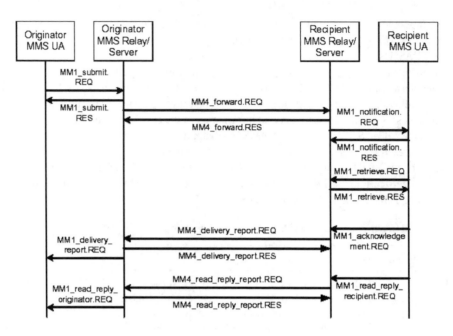

Figure 3.15   Example of sending and retrieval of an MM

Figure 3.15 shows an example of the sending and retrieval of an MM by using abstract messages. These messages can be categorized into transactions consisting of requests and responses. The labeling of these messages includes the interface through which the message is sent (e.g. MM1), the type of message (e.g. delivery) and whether the message is a request (.REQ) or a response (.RES).

Detailed description of the signaling message can be found in 3GPP specs [11]. It is important to know that the notification message from the MMS Server to the recipient User Agent can be performed via one or more SMS messages. The need for more than one SMS comes from the fact of the limited length of the SMS, and therefore for long URLs pointing to the MM or optional MMS options, several SMS may be needed leading to a higher overall delay in the notification process.

3.4.6 MMS Protocol Stack

So far, there are two possible technical realizations for the MM1 interface that have been standardized, MMS 1.0 and MMS 1.1, which are deployed on top of WAP 1.x and WAP 2.0 respectively. MMS 1.1 provides new features apart from those that were already provided by MMS 1.0. WAP 1.x is based on the WAP protocol stack (WSP, WTP, WTLS and WDP) over UDP, while WAP 2.0 also uses wireless TCP and HTTP protocols. The MMS protocol stack with WAP MMS 1.0 is depicted in Figure 3.16. The reader is kindly referred to section 3.5 for a detailed description of the WAP protocol.

As Figure 3.16 shows, the transactions between the MMS User Agent and the MMS Relay/Server always go through a WAP Gateway. These transactions have to travel over a wireless domain (MMS User Agent–WAP Gateway) and over a wired domain (WAP

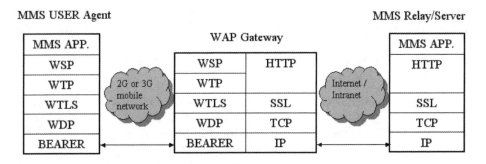

Figure 3.16   MMS protocol stack with WAP MMS 1.0

Gateway–MMS Relay/Server). It is also clear from Figure 3.16 that the transport protocols used to carry requests and responses may vary depending on the domain considered. On the wireless domain the transport protocol is performed over the Wireless Session Protocol (WSP) (in WAP 1.x) whereas on the wired domain the transport protocol is HTTP. This means that the WAP Gateway must convert HTTP requests and responses into equivalent WSP request and responses and vice versa. The target of this network configuration is the optimization of the transport of data over the wireless domain by translating requests and responses into a compact binary format and performing other transport-level optimizations. On the other hand, the usage of WAP as a transport protocol allows that many technologies can be used as bearers, as WAP separates the bearers from the MMS application layer.

Finally, at the lowest level of the protocol stack on the wireless section, several technologies can be used for the transport of data. The user subscription, the terminal capabilities and the network conditions are the main factors that influence on the transport technology to use.

3.5   WAP

3.5.1   Introduction

WAP stands for Wireless Application Protocol and actually defines not one but several protocols (or the services provided by those protocols) that inter-work together to enable the development of service applications for wireless terminals. WAP was originally defined by the WAP Forum but currently it is being further developed by the Open Mobile Alliance (OMA) [13].

The objective of WAP is to bring together telephony, wireless data and the Internet. The three technologies are growing rapidly and converging to each other. WAP intends to be the convergence point by providing a service-enabling platform for the three technologies.

Even though WAP stands for Wireless Application Protocol, it is not a protocol but a set of services, arranged in layers in a similar manner to the OSI stack. Each service can be provided by several protocols opening the door to several WAP configurations depending on the actual protocols chosen to implement each service of the WAP stack. The configuration may vary based on the actual application being enabled by the WAP stack. For instance, the configuration to access a WAP gateway can be different to the one used to access a WAP proxy. WAP also defines the content types and formats to be used, thus providing a complete platform for the development of interoperable service applications.

3.5.2 WAP Architecture

Figure 3.17 presents the WAP architecture in an example configuration. The architecture shown in Figure 3.17 reflects different functionalities (e.g. the master pull proxy and the push proxy functionalities might be built into a single proxy element).

- *Supporting servers* – provide specific functions applicable to various services (e.g. the PKI Portal that allows devices to initiate the creation of new public key certificates, a DNS server, etc.).
- *Master pull proxy* – a proxy acting as a protocol gateway, caching proxy and content encoder/decoder among other functionalities.
- *Push proxy* – a proxy that enables content providers to push information to the devices without them requesting it.
- *Application server* – is the content holder. The content may be retrieved by a pull operation or it may be delivered by the server via a push operation.
- *Secure Full Proxy* – provides access to secure networks terminating the security protocols.
- *WTA Server* – Wireless Telephony Application Server provides telephony services to the device.

More detailed information about the WAP architecture can be found in reference [14].

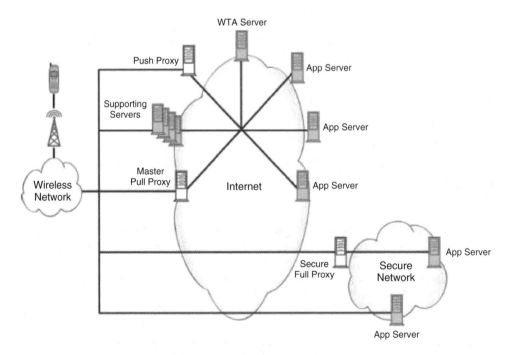

Figure 3.17   WAP architecture

3.5.3 Protocol Stack

There are currently two versions of the WAP protocol stack being used by MTs: WAP 1.x and WAP 2.x with the 'x' referring to releases within the major version.

1. WAP 1.x uses for most packet data bearers UDP as the transport protocol, providing an unreliable transmission service for the transaction – Service Access Point.
2. WAP 2.x has been able to benefit from the development in the WWW technologies to further align both WAP and the Internet. It is based on the HTTP/TCP protocol stack. The TCP implementation includes some optimizations for the wireless environment in what is known as the wireless profile of TCP (e.g. Large Window Size, Window Scale Option, Large IW, etc.). There is also the possibility of using the wireless profile of HTTP. Please refer to Chapter 7 for further details on TCP and HTTP optimization.

Figure 3.18 details the model for protocol architecture of the WAP 1.x protocol stack. The services offered at the Access Points are independent of their actual implementation, enabling

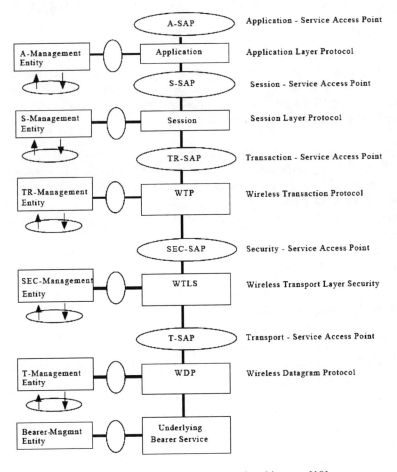

Figure 3.18   Model for the protocol architecture [10]

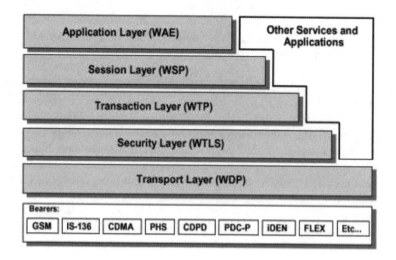

Figure 3.19 The WAP protocol stack and its interactions [10]

modifications and upgrades of the protocols without causing any changes to the interfaces used to access the services provided.

Each protocol also includes a management entity that acts as an interface between the device environment and the user. The management entity handles protocol initialization, configuration, re-configuration and error conditions that are not handled by the protocol itself. The WAP protocol stack resembles the OSI layer.

- Wireless Datagram Protocol (WDP) and Wireless Transaction Protocol (WTP) provide a similar functionality to the one found in the transport layer of the OSI stack.
- Wireless Session Protocol (WSP) provides a similar functionality to the one found in the session layer of the OSI stack.
- Wireless Application Environment (WAE) resembles the application layer of the OSI stack.

Also note that these services can be used not only by the layer above but by any other higher layer as it is clearly shown in Figure 3.19.

3.5.3.1 Wireless Datagram Protocol (WDP)

The WDP provides a datagram type of service over a wide variety of data bearers. The WDP interfaces the mobile radio bearer hiding it to higher layers of the protocol stack. WDP provides security, transaction and session management. Its main roles are to provide port number addressing, segmentation and re-assembly (SAR) and error detection. The bearer network takes care of routing and device addressing and it might also perform some SAR and error detection depending on the nature of the link underneath the bearer.

The WDP allows for multiple and simultaneous communication instances from a higher layer over a single underlying WDP bearer. The higher layer can be another protocol from the WAP stack or an application (like email). The different communication instances are identified by a port number.

The WDP supports a wide variety of bearer services. For a complete list please refer to [15]. Notice that for those bearers which use IP as the data routing protocol, WDP is implemented by UDP.

3.5.3.2 Wireless Transport Layer Security (WTLS)

The WTLS is derived from TLS [16], which in turn is derived from SSL 3.0 [17]. It is designed to work over connection-oriented and/or datagram transport protocols. Its main roles are to provide compression, encryption and authentication.

The WTLS implements security in the WAP stack and it operates on top of the transport layer protocol (Figure 3.19). The security requirements of the wireless application will determine whether the WTLS is used. The WTLS provides a transparent but secure access to the transport layer offering at the Security Service Access Point (SECSAP) an equivalent interface to the one found at the Transport Service Access Point (TSAP).

The WTLS protocol offers two main services at the SECSAP.

1. *Transport service* – for delivery of data by the WDP over the chosen bearer service.
2. *Secure connection management service* – for the establishment and termination of secure connections with a secure server.

Handshake messages used to establish a secure connection with a server might get lost when transmitted over a datagram transport protocol like WDP. The WTLS includes special mechanisms to cope with such possibilities. For a detailed description of the service primitives offered by the WTLS at the SECSAP please refer to [18].

3.5.3.3 Wireless Transaction Protocol (WTP)

The WTP provides a reliable service for a browsing environment based on requests and responses (i.e. interactive traffic). The WTP balances the reliability with the cost associated to it. WTP runs on top of a datagram service (WDP) and optionally a security service (WTLS). WTP has been defined as a lightweight transaction-oriented protocol that is suitable for implementation in 'thin' clients (MSs), and operates efficiently over wireless datagram networks.

WTP offers reliability by means of retransmissions, unique transaction identifiers, ACKs and duplicates removal. Also, to reduce the overhead on the wireless communication link, WTP does not have any explicit connection set-up and/or teardown.

The WTP offers three different classes of transaction (request and response) services – classes 0, 1 and 2:

1. *Class 0* – unreliable invoke message with no result message. It makes life easier to the WTP user so that if within an existing session, an application using WTP needs to send a datagram it does not need to connect/interface with WDP.
2. *Class 1* – reliable invoke message with no result message. It can be used for a reliable push service.
3. *Class 2* – reliable invoke message with exactly one reliable result message. Classic browsing service with request and response.

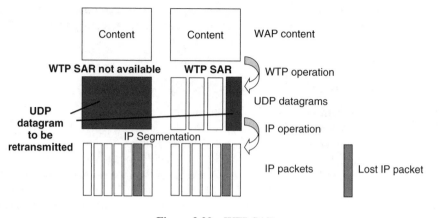

Figure 3.20   WTP SAR

One of the key issues in WAP performance is the flow-and-error control in WTP. In its simplest form, WTP provides stop-and-wait flow-and-error control, where a whole message has to be sent and acknowledge as one item.

With the WTP SAR optional feature, large WTP messages can be segmented into smaller segments, which are sent and acknowledged in groups. All the segments in a group are acknowledged by acknowledging the last packet of the group. If some segments in the group are missing when the last packet is received, then the receiver asks for selective retransmission of missing segments (Figure 3.20).

Packet groups are sent according to stop-and-wait process, meaning that only when all the packets of a group have been acknowledged, packets of a new group are sent. Optimum SAR value (i.e. packet group size) should be based on the network characteristics, packet losses and the device memory. The higher the SAR value, the higher the bandwidth utilization, i.e. less ACK messages are sent and less time is spent waiting for them. However, in case of IP packet loss, a larger amount of data must be retransmitted. For more information on the transaction services offered by WTP please refer to [19].

3.5.3.4 Wireless Session Protocol (WSP)

The session layer protocol family in the WAP architecture is called the WSP. WSP provides the upper-level application layer of WAP with a consistent interface between the two session services. The first is a connection-mode service that operates above a transaction layer protocol WTP, and the second is a connectionless service that operates above a secure or non-secure datagram transport service.

The WSPs currently offer services most suited for browsing applications. WSP provides HTTP 1.1 functionality and incorporates new features such as long-lived sessions, a common facility for data push, capability negotiation and session suspend/resume. The protocols in the WSP family are optimized for low-bandwidth bearer networks with relatively long latency. WSP provides a means for organized exchange of content between co-operating client/server applications. For more information on the session services and functionalities please refer to [20].

3.5.3.5 Wireless Application Environment (WAE)

The WAE is part of the WAP Forum's effort to specify an application framework for wireless terminals such as mobile phones, pagers and PDAs. The framework extends and leverages other WAP technologies, including WTP and WSP, as well as other Internet technologies such as XML, URLs, scripting and various media types.

The reference document [21] provides a general description of the WAE specification and acts as a root for the suite of documents that fully describe the WAE. Please refer to [21] for the complete list of WAE documentation suite. WAE has evolved dramatically from the first version WAE1 to the current version WAE2. WAE2 has been built on top of the WWW technologies. The development of the Web technologies has allowed the WAP Forum to converge and adopt more Internet standard technologies resulting in a more flexible application environment tightly paired with the WWW.

An example of such convergence is the reformulation of HTML as XML format, done by W3C and resulting in XHTML. The W3C has defined XHTML as a basic entity, XHTML Basic, which is further enhanced with different modules. The WAP Forum has used this architecture to define the XHTML Mobile Profile that extends XHTML Basic with modules, elements and attributes to provide a richer authoring language. This language is much more appropriate for resource-constrained Web clients such as mobile phones, PDAs, etc. Still, a pure XHTML Basic document is compatible with the XHTML Mobile Profile and thus, readable by WAP browsers implementing WAE2 (also known as WAP2 clients).

3.5.4 Signaling

The signaling involved in a typical pull transaction varies depending on whether the device is using WAP 1.x or WAP 2.x. The two possibilities are briefly described in the following sections, given the huge amount of terminals currently in the market that are WAP 1.x capable only.

3.5.4.1 WAP 1.x Signaling

Figure 3.21 depicts the signaling flow for a pull session using the WAP 1.x protocol stack. Note that UDP is used to provide services at the transport level – TSAP.

A connectionless datagram session is established between the WAP terminal and the WAP gateway while the gateway sets up a connection-oriented TCP-based session with the content server. Since the device does not include a DNS client, the requests for all URLs are always sent to the WAP gateway, whose address is configured in the terminal. The WAP gateway then performs the DNS query to set up the connection with the server specified by the URL.

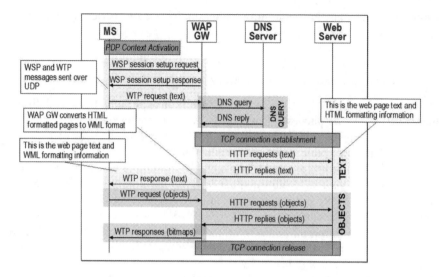

Figure 3.21   WAP 1.x example signaling flow

3.5.4.2 WAP 2.x Signaling

Figure 3.22 depicts the signaling flow for a pull session using the WAP 2.x protocol stack. Note how HTTP over TCP/IP is used to provide a connection-oriented transport service.

Still, the WAP terminal does not include a DNS server and thus the WAP gateway receives all requests just like it was seen for WAP 1.x. The main difference is the existence of

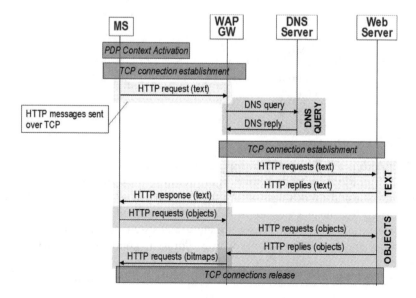

Figure 3.22   WAP 2.x example signaling flow

a TCP connection between the gateway and the terminal, including all reliability features and overheads derived from the use of TCP protocol.

3.6 Web

3.6.1 Introduction

The term 'Web' is normally used to refer to the millions of HTML pages that Web servers host around the world. It has quickly changed the way people see the Internet, putting aside on its rampage to success several other old brothers like Gopher or Archie. However, it is not the purpose of this section to explain the keys for the success of the WWW and the effects that it has on the usage and explosion of Internet. This section will briefly explain the internals of the service with the idea of setting the foundations for later understanding of how the Web experience can be fine-tuned for an optimum mobile experience.

The Web is, together with the email, the main service used over the Internet nowadays. Thus, it is obvious that the mobile industry shall deliver a rich Web experience if they want mobile users to become greedy data users. The task is not easy as most Web pages assume a minimum screen resolution of 800×600 pixels while most mobile phones do not go over 176×208 pixels. Only certain models designed for the ultimate mobile task force offer resolution as high as 640×320 pixels, with a high price to be paid in terms of size of the terminal.

Several start-up companies have taken the opportunity to create software that improve the rendering of Web pages in small screens trying to make the Web more suitable for the average terminal. Still, due to these usability limitations, most mobile users browsing the Web do not do it with a Web client running in their terminal but with their favorite Web client running on a laptop and using the mobile phone as a modem to gain access to the Internet.

3.6.2 Architecture

The Web architecture uses a client–server approach in which the exchange of information is done with transactions: requests and responses. The user types an URL Web address and the Web browser contacts the Web server, submits the request for the URL and waits for the response. Typically the response will be an HTML page that normally includes references to other URLs that also need to be requested for the page to be properly displayed. The browser will then automatically request all these embedded URLs in subsequent transactions.

The HTML page sent by the Web server can be static, i.e. all content has been created prior to the request, or it can be dynamic, i.e. part (if not all) of the content of the Web page is created *ad hoc* for this request in particular. The degree of particularization varies and the methods for creating dynamic content are continuously evolving together with the technology. Initially the Web clients were designed as dummy clients, with all the power concentrating on the server. This approach fitted well the beginnings of the WWW, when most of the content was static and the server had to execute very little processing to serve the Web page requested.

Now, much of the content of a Web page is dynamic. This calls for a distributed execution of the processing required for the generation of the Web page, otherwise the servers would be put under a heavy workload that would negatively affect its performance. That is how Web browsers are now turning smarter every day and users can add new functionalities to them via plug-ins that enable rendering of even more complex (and not fully standardized) content at the client side.

The Web is built on top of three corner stones.

1. *HTTP*: The HyperText Transfer Protocol is an application protocol that runs on top of TCP/IP and provides the transaction service between client and server. Currently, HTTP 1.1 is the most extended and better performing version, although still compatible with older version HTTP 1.0.
2. *HTML*: The HyperText Markup Language provides the syntax for representing Web pages. The HTML has evolved dramatically since its creation in 1990 and after four main revisions it has now been reformulated in XML with the release of XHTML 1.0. The reader is kindly referred to [22] and [23] for further information on HTML, its current status and evolution.
3. *URI*: The Uniform Resource Identifier provides schemes for addressing resources on the net. The most commonly used Web browsing scheme is 'http:'. For more information on URIs please see [24] and [25].

On top of these three elements, developers have been adding extra tools to enable the creation of more powerful and richer hypertext content. Such tools include CGI servers, Java and JavaScript, and many others. The WWW itself includes the most up-to-date information about the Web architecture and its enablers. The World Wide Web Consortium [26] has worked to progress in the definition of new Web architectures. The latest edition of the Recommendation 'Architecture of the World Wide Web Volume One', can be found in [27].

Another key element of the Web architecture is the DNS server. The DNS server resolves an URI of the type 'http://www.wiley.com' into an IP address that the Web browser can use to address the request to the right Web server.

3.6.3 Protocol Stack

The basic protocol stack used by mobile Web browsers is based on HTTP/TCP/IP, which is briefly described in the following subsections.

3.6.3.1 HyperText Markup Language (HTTP)

HTTP has already been introduced earlier. It is a simple network protocol that provides the transaction services needed by Web browsers. The main resource for HTTP information is the W3C [28]. The W3C has closed its activities on HTTP after standardizing HTTP 1.1 [30]. This version of HTTP overcomes many of the problems that HTTP 1.0 [29] encountered when Web browsers went wireless. Main HTTP1.1 enhancements compared to 1.0 version are listed below [31].

Table 3.5 HTTP 1.0 vs HTTP 1.1

HTTP version	Description
HTTP 1.0	Requires the establishment of a new TCP connection per object included in the Web page.[a]
HTTP 1.1 (*Persistent connection*)	Allows reusing the same TCP connection for the subsequent objects.
HTTP 1.1 (*Persistent connection + Pipelining*)	In addition to the persistent TCP connection (explained before), different object requests can be sent simultaneously by the terminal without waiting for the reception of the previous one from the server.

[a] Depending on the browser implementation, several connections in parallel may be activated to improve the performance.

- *Persistent connection feature* (mandatory) – allows reusing the same TCP connection for subsequent objects included in the Web page.
- *Pipelining feature* (optional) – the different object requests can be sent simultaneously without waiting the reception of the previous one.

Table 3.5 describes the main differences between HTTP 1.0 and 1.1.

The W3C is now working on an XML protocol in line with the HTML re-orientation toward XML with XHTML 1.0. The W3C has promised to also specify a binding between HTTP and the new XML protocol to ensure backward compatibility.

3.6.4 Signaling

The signaling involved in a Web browser transaction is simple, in line with the design concepts behind the HTML language and the HTTP protocol.

As it was mentioned in section 3.6.2, when a user types a resource location using a URI in the Web browser, the following actions take place (Figure 3.23).

- The user types a URI (e.g. http://www.wiley.com).
- The Web browser sends a DNS request to a DNS server to obtain an IP address associated with the Web server www.wiley.com.
- The DNS replies to the request from the Web browser with a response message containing the IP address associated with the Web server www.wiley.com.
- The Web browser sends an HTTP request to the IP address of the Web server www.wiley.com obtained through the DNS server.
- The Web server receives the HTTP request and executes whatever internal processing is required to serve the Web page.
- The Web server replies to the Web browser with an HTTP response message containing the HTML content of the Web page www.wiley.com.
- The Web browser receives the HTML data within the HTTP response and renders it on the screen. Web pages typically include both text and objects (e.g. pictures, tables, scripts, etc.), whose URL are defined in the HTML code.

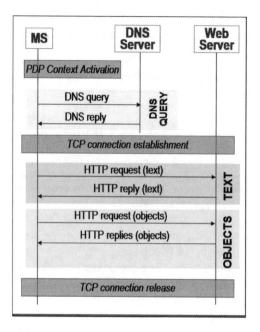

Figure 3.23   Signaling example for a Web request

- Depending on the number of embedded objects in the Web page and the HTTP version in use, a different number of TCP connections are established with the respective servers to retrieve aforementioned objects. Typically, HTTP 1.1 establishes two or four parallel TCP connections in addition to the persistent connection feature.

These steps are presented in Figure 3.23. As mentioned earlier, the initial HTML content might include references to further URIs required to complete the Web page. The Web browser will issue new HTTP request messages to retrieve all required resources.

3.7 Push-to-Talk over Cellular (PoC)

3.7.1 Introduction

PoC is a new service that introduces a direct one-to-one and one-to-many voice communication service in cellular networks. From functionality point of view, PoC is a 'walkie-talkie' like service, since it is based on half-duplex communication and the user needs to reserve the channel before transmitting the speech. PoC service can be mainly used for two different types of communication (one-to-one PoC call and group PoC call).

1. *One-to-one PoC communication*: A voice communication is taking place between two users, where only one user can speak at a time. Each user takes control of the communication by pressing one button (PTT button). There are two subclasses of calls depending on the called party acceptance mode.

(i) *One-to-one direct call*: No called party action is required.

(ii) *Instant personal alert call*: A PoC client requests another PoC subscriber to initiate a one-to-one session.

2. *Group PoC communication*: One user has a voice communication with a multiple number of users, where users talk one at a time. There are three types depending on group communication set-up.

    (i) *Pre-arranged group session*: Group members are predefined. The session starts when any member of the group invites other members to join the session. Only members of that predefined group can join.

    (ii) Ad hoc *group session*: One user invites other users to participate in a group call, without the need to be members of a common group. It is a temporary group created only for that session.

    (iii) *Chat group session*: A user joins a chat group to participate in a group communication. Access to chat groups may be restricted or unrestricted.

Current voice service is relayed over CS bearers, which means that the resources are established and maintained during the whole call duration and are released only when terminating the call (i.e. full duplex communication). The full duplex communication has very strict delay requirements in order to achieve a good end-user perception. A reasonable maximum end-to-end delay value for voice calls could be around 400 ms.

Optimization of spectral efficiency and the stringent delay requirements make recommendable the use of packet data bearers for PoC service. When using packet data bearers, the resources are not reserved on both directions, but only in one direction and when required (based on voice activity). Figure 3.24 shows an example of airtime usage between a full duplex call and a PoC call.

The half-duplex nature of the service relaxes the delay requirements, which are not as strict as in conventional voice calls. When using the service, the user has to first press the PTT button to access the channel, and then start to talk when hearing an indication confirming that the channel is granted.

Figure 3.25 describes the difference between CS call and PoC call from timeslot usage point of view for the case of (E)GPRS network. While in CS, one timeslot, or half depending

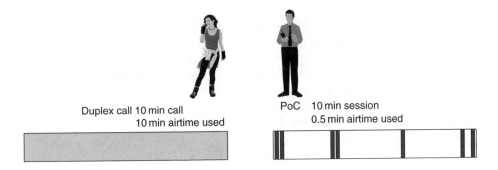

Duplex call 10 min call
10 min airtime used

PoC   10 min session
0.5 min airtime used

Figure 3.24   Full duplex CS call vs half-duplex PS PoC call [35]

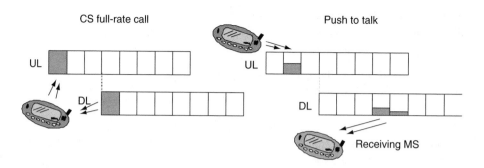

Figure 3.25   Resource allocation comparison between CS call and PS PoC call [35]

on full-rate or half-rate utilization, is reserved in the UL and in the DL, and with PoC over PS only one direction is reserved and only when there is user data to be transmitted.

3.7.2 PoC Architecture

This section introduces the PTT functional architecture as it is described by OMA. It is based on IMS as specified in [32]. This solution has two end points: the UE in one side and the PoC server in the other side. All information, both voice and signaling, flow between these two entities. In Figure 3.26, the functional architecture is outlined according to OMA architecture [48]:

The access in the PoC architecture includes both the radio access and all the nodes required to gain IP connectivity (e.g. GGSN). The PoC architecture contains two major functional entities.

1. *PoC client* – resides on the mobile terminal and is used to access PoC service.
2. *PoC server* – implements the application level network functionality for the PoC service. The PoC server may perform a *Controlling PoC Function* or *Participating PoC Function*.

As part of the *Controlling PoC Function*, the PoC server needs to perform a series of functions to support and maintain a session. The server keeps track of all the signaling procedures, synchronization and user handling. It:

- Provides centralized PoC session handling
- Provides the centralized media distribution
- Provides the centralized talk burst arbitration functionality including talker identification
- Provides SIP session handling, such as SIP session origination, termination, etc.
- Provides policy enforcement for participation in group sessions
- Provides the participants information
- Collects and provides centralized media quality information
- Provides centralized charging reports
- Supports user plane adaptation procedures
- May provide transcoding between different codecs
- Supports talk burst control protocol negotiation.

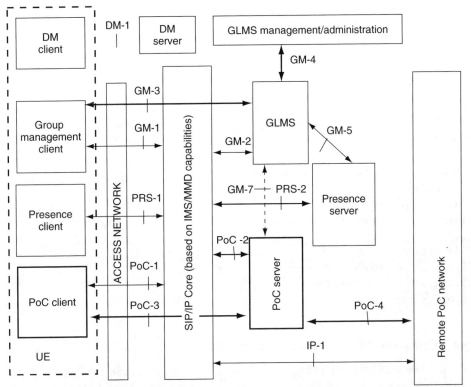

1. Bold box identifies PoC functional entities.
2. Remote PoC Network contains the same network elements and reference points as the home PoC network.

Figure 3.26   PoC architecture diagram [48]

In the *Participating PoC Function*, the PoC server needs to provide functions to manage user plane of the connections: session control, access control, charging, etc. It:

- Provides PoC session handling
- Supports the user plane adaptation procedures
- May provide the talk burst control message relay function between PoC client and controlling PoC server
- Provides SIP session handling, such as SIP session origination, termination, etc., on behalf of the represented PoC Client
- Provides policy enforcement for incoming PoC session (e.g. access control, incoming PoC session barring, availability status, etc.)
- Provides the participant charging reports
- Supports talk burst control protocol negotiation
- Stores the current answer mode and incoming PoC session barring preferences of the PoC client.

Apart from these two major entities in the system, PoC client and PoC server, there are other entities which provide external services to the PoC system. These entities are aimed to provide support for user management, charging, group maintenance, etc.

3.7.2.1 SIP/IP Core

The SIP/IP core includes a number of SIP proxies and SIP registers (see also section 3.7.3 for more information on SIP). The SIP/IP core performs the following functions, which are needed to support the PoC service.

- Routes the SIP signaling between the PoC client and the PoC server
- Provides discovery and address resolution services
- Supports SIP compression
- Performs authentication and authorization of the PoC user at the PoC client based on the PoC user's service profile
- Maintains the registration state
- Provides support for identity and privacy on the control plane
- Provides charging information
- Provides capabilities to Lawful Interception.

3.7.2.2 Group and List Management Server (GLMS)

PoC users use the GLMS to manage groups and lists (e.g. contact and access lists) that are needed to access to the multicast services supported by PoC. The GLMS performs the following functions:

- Provides list management operations to create, modify, retrieve and delete groups and lists for authorized users
- Provides storage for groups and lists
- Provides Notifications of modifications to lists.

3.7.2.3 Presence Server

The presence server is in charge of tracking the presence of users in the system and keeps track of the location where they can be found, when data is sent to them. It:

- Maintains the presence status of PoC users
- Supports the publication of presence information on PoC users from other PoC users
- Supports the watching and fetching of presence information on PoC clients by other PoC clients
- Supports the publication of PoC subscribers specific presence information (based on the presence information provided by the participating PoC function, the specific availability status of a PoC client is communicated to other PoC clients)
- Allows the PoC server, on behalf of the PoC client, to watch and fetch presence information, i.e. PoC-specific and general presence information

- Supports the authorization of watchers of PoC clients presence information and authorizes the watching and fetching of presence information
- Supports the authorization of presence list subscriptions
- Supports back-end subscriptions of presence lists containing members in other domains
- Supports the watching and fetching of presence information from other presence servers (presence list).

3.7.2.4 Charging Entity

This is an external entity, which may reside in the operator's domain. This entity takes various roles, when network operators and/or service providers need to perform the charging activities.

PoC charging architecture supports both subscription and traffic-based charging. For the subscription-based charging, subscription events like service activation time and subscriber PoC service profile are provided to the charging infrastructure. For traffic-based charging, data is provided to the charging infrastructure in time to support both pre-paid and post-paid billing models.

3.7.2.5 Device Management Server

The device management server is in charge of controlling the interaction between the different devices and the PoC server, and maintains software compatibility between the handsets. It:

- Initializes and updates all the configuration parameters necessary for the PoC client
- Supports software update for application upgradeable handsets. The PoC service provider shall be able to set up PoC communication feature configurations remotely in the terminal device by using the device management mechanism. The PoC service provider shall also be able to update PoC communication feature configurations remotely in the terminal. The PoC UE shall be able to receive the contents sent by service provider.

Functionality wise, the process followed when using the service is the next one: when the calling party wants to initiate a PoC call, the user presses the PTT button. Then, this will originate the corresponding user permission checking in the PoC server. Then, after those checks are verified, the calling party (UE) would receive the permission from the server in the form of an audible beep. The user may then start to talk and IP packets are generated, which are transmitted through the access network toward the PoC server (IMS core and PoC server). The server makes the required access, checking and forwarding of packets toward the recipient(s) or called parties.

3.7.3 PoC Protocol Stack

PoC service has adopted a protocol-stack based on RTP/RTCP protocols for the user plane and SDP/SIP for the control plane [36] (Figure 3.27).

The following protocols are used above transport layer.

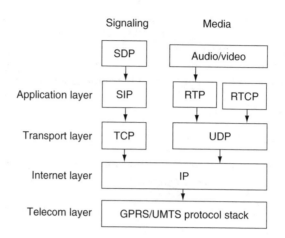

Figure 3.27 PoC protocol stack

- *Session Initiation Protocol (SIP)* – is a client–server protocol to control delivery of data. It is used for registration, session set-up and session renegotiation. It does not convey media streams itself. SIP protocol is defined in [41]. SIP signaling may be compressed on the radio interface using [43–46].
- *Session Description Protocol (SDP)* [42] – conveys information on the media capabilities of the PoC client. SDP includes information during the SIP negotiation about the media encoding and port numbers used for the media streams.
- *Real-Time Protocol (RTP)* [38, 39] – is used for media transport. RTP carries data with real time requirements. RTP flow can be compressed with Robust Header Compression (ROHC) scheme according to [38]. There is a specific payload type to transmit voice using Adaptive Multi-rate (AMR) codecs [40].
- *Real-Time Control Protocol (RTCP)* – conveys participants information and monitors the quality of the RTP session (quality feedback).

PoC service makes use of AMR speech codecs (see also section 8.4.2 for further details on AMR codec) in order to provide efficient voice codification. Terminals shall support AMR 5.15 as mandatory and default codec, but it is also desirable to use the support of AMR 4.75. During the call set-up phase (SIP set-up), both terminal and PoC servers should be able to support codec negotiation, according to operator's configuration. Codec mode adaptation is supported in PoC service, although it shall not be done during a talk burst. The coded AMR frames are packed into RTP packets following a standardized payload format.

It is recommendable that PoC service uses two parallel PDP contexts according to the following criteria: if streaming traffic class is supported, then interactive class PDP context is recommended for control signaling (SIP and HTTP) and streaming class PDP context for media information (RTP and RTCP). If streaming traffic class is not available, then two interactive PDP contexts, one for signaling and the other for media information; or a single interactive PDP context, for both signaling and media information, should be used.

PDP contexts are established when user connects to PoC service. Therefore, always-on PDP context is also recommended to optimize call set-up time and avoid that every time the PTT button is pressed, the user has to wait for PDP context establishment.

3.7.4 PoC Signaling

This section shows an example of the signaling sequences of the establishment of one *ad hoc* or one-to-one PoC session using on-demand signaling in the originating side. As a prerequisite for the on-demand signaling, the PoC client needs to be registered in the service. Figure 3.28 shows the signaling sequence for the manual answer case [47].

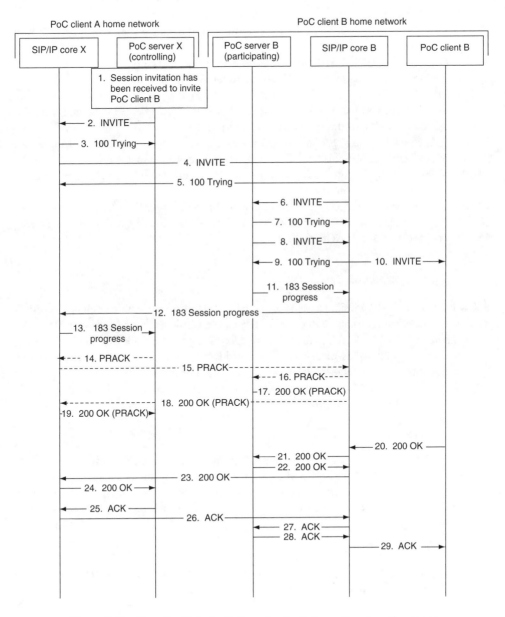

Figure 3.28   Signaling flow for PoC session invitation with automatic answer

The different steps shown in Figure 3.28 are described as follows.

1. PoC server X, which in this example performs the controlling function, is triggered to send INVITE request to PoC client B.
2. PoC server X sends the INVITE request message to the SIP/IP core X, which is the entity in charge of the transmission of the message.
3. SIP/IP core X sends 100 Trying response to the PoC server X.
4. SIP/IP core X forwards the INVITE request to the SIP/IP core B.
5. SIP/IP core B sends 100 Trying response to SIP/IP core X.
6. SIP/IP core B forwards the INVITE request to the PoC server B, which performs as participating function.
7. PoC server B sends 100 Trying response to the SIP/IP core B.
8. PoC server B sends the INVITE request to the SIP/IP core B.
9. SIP/IP core B sends 100 Trying response to the PoC server B.
10. SIP/IP core B forwards the INVITE request to the PoC client B. The end-user is at this time notified of the PoC session invitation.
11–13. PoC client B sends 183 Session progress response via signaling path. In this case of automatic answer, this Session progress response is communicated to PoC server X.
14–16. If PoC server B requested 183 response to be sent reliably, PoC server X sends PRACK request to the PoC server B.
17–19. PoC server B sends 200 OK for PRACK to the PoC server X.
20–24. PoC client B sends 200 OK response for INVITE via signaling path to PoC server X.
25–29. PoC server X sends ACK request.

3.7.5 PoC Performance Requirements

PoC service is especially sensible to delay and quality. From user perception point of view, these are the two main requirements and in general delay can be considered as the main one. Hence it is quite important to control the different sources of delay that may affect a PoC communication.

There are two main kinds of delays to be considered, depending on the part of the PoC communication that is affected. Delays during communication setup, which affect to the initiation of one call, and delays during communication, which affect the latency that the receiver end will notice in the communication [37].

1. *Delays in communication set-up*:

   - *Right-To-Speak (RTS) delay* – is defined as the time between the instant a PoC subscriber initiates a PoC session and when it receives a 'right-to-speak' indication.
   - *Start-To-Speak (STS) delay* – is defined as the time between the instant a PoC participant initiates a floor request on an ongoing session and when it receives a 'start-to-speak' indication.

2. *Delays during the communication*:

   - *Voice Delay Time (VDT)* – is defined as the time since message is spoken by a calling party until it is heard by the invited called party.
   - *Round-Trip Time (RTT) delay* – is defined as the time since a message is sent from the calling party, till an answer is received from the remote end.

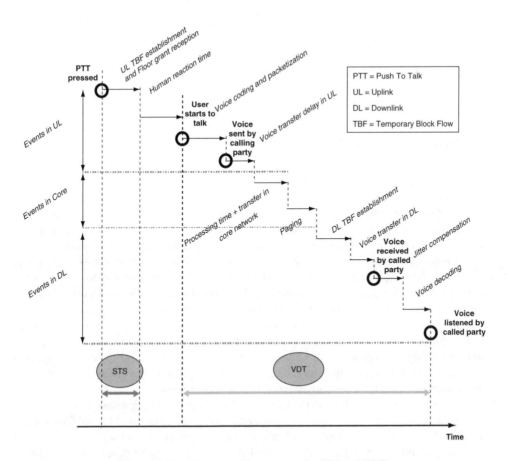

Figure 3.29   Diagram with processes for STS and VDT

Figure 3.29 shows different processes taking place during a talk burst communication.

Based on the definition of these basic performance measurements it is possible to define the performance requirements, in terms of delay, that shall be fulfilled by a PoC communication in order to provide an acceptable service level to all the participants. The performance requirements have to be fulfilled up to 10% BLER conditions. The values are not absolute maximum values, but expected to be fulfilled by a majority of users in a well-designed network.

- *Start-to-speak delay* – from user pressing PTT button until receiving permission to talk. This indicator is very important for end-user perception, since the PoC service users do not want to wait for a long time before being able to initiate the conversation. It is recommended for this delay to be below 1.6 seconds.
- *Voice delay time* – since message is spoken until it is heard by receiving user. Again, this indicator is very important for the end-user perception. A service with long communication delays would not allow for an interactive conversation. This delay should typically be no more than 1.6 seconds during the session. However, for the first talk-burst, and in the case that it also includes the session set-up, a voice delay of up to 4 seconds is acceptable.

- *A user should be able to join a group* – hear or initiate a talk burst, within 4 seconds after the initiation of the join procedure.
- *Voice quality* – should meet the requirement of an MOS above 3 at BER below 2%. Voice quality is one of the most important indicators. If the voice quality is not good enough, the service is unacceptable. If we assume we have no packet loss, voice quality will be determined by the codec we use for the PoC data.

3.8 Network Gaming Services

3.8.1 Introduction

The gaming and mobile phone industries have recently discovered the phones as a new powerful gaming platform. The increasing processing capacity of the terminals, bigger color screens, big memories and extremely high-connectivity capabilities have made the MT a great platform for developing new games or adapting well-known and very popular games. For the gaming industry, mobile phones represent a new market. For mobile operators, games, and especially games requiring interactive communications, are a powerful way to increase their Average Revenue Per User (ARPU). They might charge the user for downloading the game (game fee) plus for the data download. Mobile phone manufacturers and operators, are also using the games as a driver to push users to upgrade their 'old' phones to more powerful ones.

Furthermore, the games can also generate new data usage. For instance, single-player games might upload high scores to a gaming server so that users can compete for the highest score. However, for mobile operators, interactive multi-player games bring the highest data usage and at the same time the highest performance requirements.

There are many factors affecting the success of the network gaming industry, like platform fragmentation, network requirements, interoperable APIs, etc. Ericsson, Motorola, Nokia and Siemens all founded in July 2001 the Mobile Game Interoperability Forum [33] trying to set an industry standard for a gaming platform that would allow game developers to produce games that can be executed in different terminals (all compatible with the MGIF-compliant gaming platform). The goal of the MGIF was to define an API, not to implement the gaming platform itself.

The work of the MGIF was consolidated into the OMA [13] in February 2003. The OMA is continuing with the work started in MGIF and ensuring that its specifications are fully coordinated with other ongoing work in the OMA that could be of relevance for the success of the network gaming services like presence services, instant messaging, etc.

All these efforts concentrate on providing an interoperable platform that allows game developers and gaming platform developers to reach the gaming market (and do it successfully) with reduced costs, thanks to the interoperability capabilities. However, none of those issues will be covered in this section. The reader is kindly referred to the OMA Web page where he/she will find a comprehensive list of resources related to the service-enabling work. This section will introduce the reader to the impact that the network performance has on the mobile gaming performance and identify those points where special care should be taken to ensure a successful gaming experience.

From the network requirements point of view, games can be classified into different categories. First of all, there are peer-to-peer games and server-based games. Peer-to-peer games are most commonly two-player games in which the players exchange information about its own status within the game. A peer-to-peer game could include more than two players. However, the approach has some scalability problems as each MT needs to keep a connection with the other players and while that might be feasible if there are three other players on the game, it might pose a problem (in terms of memory, processing power, OS capabilities, etc.) if a terminal needs to keep up and attend to several dozens of connections. Server-based games tackle this problem by moving the scalability issues to a workstation located behind the mobile network. Thus, the mobiles only need to keep up with one single connection to the server. The requirements from the network perspective are more or less the same regardless of whether the game is being played as peer-to-peer or server-based.

Another possible classification is based on the behavior and interactivity of the user with the game and with other players.

- *Action games* – games of the type *shoot 'em all*, racing games and, in general, games that require a quick reaction of the player to the events ongoing within the game (normally generated by other players) in order for him/her to succeed or even remain alive in the game. These games impose the most stringent requirements on the networks for a satisfactory gaming experience.
- *Real-time strategy games* – these games involve several players gaming simultaneously. Every player shall be notified promptly about the changes in the status of the game due to the actions of other players. Still, the real-time requirements are not so hard as with action games.
- *Turn-based games* – these games also involve several players but there is a clear order in which each of them will generate events within the game. A typical example of a turn-based game could be chess. These games are characterized by the fact that the player will use some time executing its actions before the turn is given to a different player. The other parties wait patiently for the player with the turn to execute its moves and thus the requirements for the networks are not so strict.

3.8.2 Network Requirements

The first step to set the network requirements is to identify the gaming requirements. Mobile games typically will use TCP for session set-up and control with the gaming server, and UDP for conveying information about any event triggered by a player. For every event, a UDP packet can be sent setting a traffic profile of many small UDP packets that need to reach its destination with as little delay as possible. In this case, order of arrival is also important as only one event might be possible because another one was triggered before. UDP does not guarantee ordered delivery and so the protocol used by the game (e.g. the communication protocol used by the gaming platform) will have to take that responsibility. Still, the benefits of using UDP over TCP in terms of delay are enough to justify such approach.

Each game is different and it would be impossible for mobile operators to optimize their network performance to maximize the gaming experience for every mobile game in the market. But there is one requirement that affects in general to any game and that is important enough to dramatically improve on its own the gaming-performance experience: *delay*.

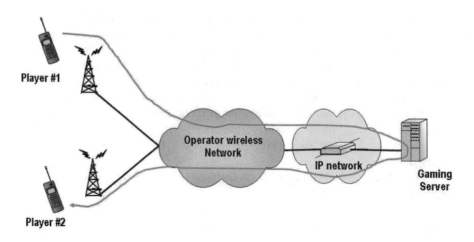

Figure 3.30   Network gaming scenario

A gaming network is a network that has a low RTT. The following delays can be identified when analyzing the evolution of a packet[1] (Figure 3.30).

- *Delay from the mobile phone to the edge of the mobile network* – delay of crossing the mobile network.
- *Delay from the edge of the network to the gaming server* – this delay is set by the number of routers that the packet needs to go over through the IP network in order to reach the gaming server.
- *Delay introduced by the processing done by the gaming server* – this delay is game specific and is out of the control of the mobile operator.
- *Delay from the gaming server back to the edge of the mobile network* – this can be considered roughly the same as the delay in the other direction (delay #2).
- *Delay through the mobile network* – delay to the mobile phone which receives the packet.

There is clearly one point where mobile operators can act easily and that is the location of the gaming server. If the gaming server is within the mobile operators' IP network, the delays due to 2 and 4 can be roughly dismissed. The mobile operator has no control over the delay introduced by this concept when the gaming server is on the Internet. Moreover, proxies and catching systems will not help here as each packet is unique. Having the gaming server within the mobile operators' network also has commercial implications since only subscribers of that operator will be able to access it. Is this good or bad? It depends on the business model that the operator wants to use.

The critical point where mobile operators, network manufacturers and standardization bodies shall concentrate to improve the gaming experience of mobile players is on the delays introduced by the mobile network itself. As shown above, the RTT of a small packet (ca. 100

---

[1] Assuming two players using mobile platforms.

bytes) is the critical factor to optimize. Packets are small enough that the bit rate not to be a relevant factor. The games send little information but when they need to do so then it shall go fast. The latency of the network is the key to a successful mobile-gaming experience!

References

[1] 3GPP TS 23.125, Overall high level functionality and architecture impacts of flow based charging, Stage 2, V6.2.0.

[2] 3GPP TS 23.228, IP Multimedia SubSystem (IMS), Stage 2, V6.7.0.

[3] S. Floyd and T. Henderson, 'The NewReno Modification to TCP's Fast Recovery Algorithm', RFC 2582, April 1999.

[4] M. Allman, V. Paxson and W.Stevens, 'TCP Congestion Control', RFC 2581, April 1999.

[5] M. Allman, S. Floyd and C. Partridge, 'Increasing TCP's Initial Window', RFC 3390, October 2002.

[6] V. Paxon, 'Computing TCP's Retransmission Timer', RFC 2988.

[7] Technical Spec. 23.038, Alphabets and Language-specific Information, 3rd Generation Partnership Project.

[8] ISO/IEC 10646-1, Universal Multiple-Octet Codec Character Set (UCS), 1993.

[9] 3GPP TS 23.040, 'Technical realization of the Short Message Service (SMS)' (Release 4), V4.7.0, June 2002.

[10] WAP Forum Web page, http://www.wapforum.org.

[11] 3GPP TS 23.140, Multimedia Messaging Service (MMS); Functional description; Stage 2 (Release 6), V6.7.0, September 2004.

[12] 3GPP TS 23.057: 'Mobile Execution Environment (MExE); Functional description, Stage 2'.

[13] OMA Open Mobile Alliance, http://www.openmobilealliance.org.

[14] WAP Architecture Specification, WAP-210-WAPArch.-20010712.

[15] Wireless Datagram Protocol Specification, WAP-259-WDP-20010614-a.

[16] T. Dierks and C. Allen, 'The TLS Protocol' January 1999, ftp://ftp.isi.edu/in-notes/rfc2246.txt.

[17] A. Frier, P. Karlton and P. Kocher, 'The SSL 3.0 Protocol', Netscape Communications Corp., November 18, 1996.

[18] Wireless Transport Layer Security Specification, WAP-261-WTLS-20010406-a.

[19] Wireless Transaction Protocol Specification, WAP-224-WTP-20010710-a.

[20] Wireless Session Protocol Specification, WAP-230-WSP, approved version 5 July 2001.

[21] Wireless Application Environment Specification, WAP-236-WAESpec-20020207-a.

[22] Hypertext Markup Language (HTML) Home Page, http://www.w3c.org/MarkUp/.

[23] HTML recommendations http://www.w3c.org/MarkUp/#recommendations.

[24] Naming and addressing: http://www.w3c.org/Addressing/.

[25] URIs, URLs and URNs: Clarifications and Recommendations 1.0: http://www.w3c.org/TR/uri-clarification/.

[26] World Wide Web Consortium, http://www.w3.org/.

[27] Architecture of the World Wide Web, 1st edition, http://www.w3.org/TR/webarch/.

[28] Hypertext Transfer Protocol http://www.w3.org/Protocols/.

[29] T. Berners-Lee, R. Fielding and H. Frystyk, Hypertext Transfer Protocol – HTTP/1.0, RFC 1945, May 1996.

[30] R. Fielding, J. Gettys, J. Mogul, H. Frystyk, L. Masinter, P. Leach and T. Berners-Lee, Hypertext Transfer Protocol – HTTP/1.1, RFC 2616, June 1999.

[31] G. Gómez, J. Torreblanca and J. Ramiro-Moreno, 'Assessing and Optimizing End User Performance over 2G/3G Cellular Networks', IASTED CSN'2004, Marbella (Spain).

[32] 3GPP TS 24.229, Internet Protocol (IP) multimedia call control protocol based on Session Initiation Protocol (SIP) and Session Description Protocol (SDP), Stage 3, Release 5.

[33] MGIF, Mobile Game Interoperability Forum.

[34] H. Inamura, G. Montenegro, R. Ludwig, A. Gurtov and F. Khafizov, 'TCP over Second (2.5G) and Third (3G) Generation Wireless Networks', RFC 3481.

[35] T. Halonen, J. Romero and J. Melero, 'GSM, GPRS and EDGE Performance', Ed. Wiley, 2nd edition, 2003.

[36] OMA PoC User Plane, Draft Version 1.0.9, October 2004.

[37] OMA Push to Talk over Cellular Requirements, Draft Version 1.0, June 2004.

[38] Carsten Bormann, et al., 'Robust Header Compression (ROHC)', RFC 3095, July 2001.

[39] IETF RFC 1889, 'RTP: A Transport Protocol for Real-Time Applications'.

[40] IETF RFC 3267, Real-Time Transport Protocol (RTP) Payload Format and File Storage Format for the Adaptive Multi-Rate (AMR) and Adaptive Multi-Rate Wideband (AMR-WB) Audio Codecs.

[41] IETF RFC 3261, 'SIP: Session Initiation Protocol'.
[42] IETF RFC 2327, 'SDP: Session Description Protocol'.
[43] IETF RFC 3320, 'Signaling Compression (SigComp)'.
[44] IETF RFC 3321, 'Signaling Compression (SigComp) – Extended Operations'.
[45] IETF RFC 3485, 'The Session Initiation Protocol (SIP) and Session Description Protocol (SDP) Static Dictionary for Signaling Compression (SigComp)'.
[46] IETF RFC 3486, 'Compressing the Session Initiation Protocol (SIP)'.
[47] OMA PoC Control Plane, Draft Version 1.0, 4 September 2004.
[48] OMA Push to talk over Cellular (PoC) – Architecture, Draft Version 1.0, 7 October 2004.
[49] Pablo Ameigeiras, Jeroen Wigard and Preben Mogensen, 'Impact of TCP Flow Control on the Radio Resource Management of WCDMA Networks' Vehicular Technology Conference, VTC, Spring 2002.

4

Quality of Service Mechanisms

Raquel Rodríguez, Daniel Fernández, Héctor Montes, Salvador Hierrezuelo
and Gerardo Gómez

4.1 What is Quality of Service?

Some years ago, data telecommunication networks caused an important revolution worldwide, allowing quick and easy communication between parties. At that stage, very basic services like file transfer or e-mail were supported, and service requirements were not an issue. However, one of the most significant achievements along the years has been the support for a richer variety of services and a higher level of service personalization. The different nature of the services implies particular quality requirements that need to be fulfilled in order to satisfy the end-user expectations. For instance, listening to the radio through the Internet requires a lower delay compared to an e-mail service, while the e-mail service requires higher reliability compared to the radio audio information.

This basic service classification is just the starting point for different research areas and the development of a wide set of mechanisms and protocols oriented to classify, differentiate or provide a special treatment to each data flow. These mechanisms are the principles of *Quality of Service* (QoS) management.

This chapter is aimed to provide an overview of what QoS stands for, describe which kind of mechanisms are currently available in IP and cellular networks, introduce the need for QoS differentiation and provide a classification of data services according to the QoS requirements.

4.1.1 QoS Definition

The term 'QoS' can be defined as the system's ability to provide a selective treatment to different services or users in the most cost-efficient way. A wide collection of technologies, including different radio access, and techniques can be used in order to provide predictable results.

End-to-End Quality of Service over Cellular Networks: Data Services Performance and Optimization in 2G/3G
Edited by G. Gómez and R. Sánchez © 2005 John Wiley & Sons, Ltd

The final goal of the QoS is to provide a good end-user experience when using a particular service. Considering that the end-users' perception of the 'quality of a service' is different depending on the application requirements, it is important to define a set of common parameters which allow expressing the service requirements to the underlying protocols.

QoS can be described and applied in a qualitative or quantitative manner. A qualitative QoS management is based on a relative treatment applied to a certain traffic type compared to other traffic types. For instance, giving higher priority to a Web traffic compared to e-mail may be used by the network to provide a relative faster forwarding of the Web traffic, although no absolutes guarantees are provided. On the other hand, a quantitative QoS management tries to guarantee certain transmission capabilities required by the service, and relies on the definition of metrics such as throughput, delay or loss. For instance, a video-streaming application may require a guaranteed bit rate of 32 kbps in order to be reproduced in the terminal without any interruption and with acceptable quality. A quantitative QoS negotiation is also called a Service Level Agreement (SLA).

It is important to understand that the quantitative QoS management generally requires additional and more complex functions than the qualitative QoS management. By implementing features like Admission Control (AC) or Resource Reservation (RR), the network is able to ensure a certain service level at the expense of blocking other flows. At this point, there is a trade off between service quality and system capacity in terms of number of users served. For that reason, systems implementing quantitative QoS require a more complex dimensioning exercise during the network deployment, which is critical to support and satisfy the traffic needs.

The network should also be responsible for ensuring that resources being used by a particular flow are in accordance with the negotiated QoS. Policing functions may trigger some actions to react against non-compliant traffic (e.g. traffic shaping).

4.1.2 Need for QoS Differentiation

Internet has been providing a Best Effort (BE) service, giving the same treatment to all data services. This approach has not been an issue during last years, and BE networks are already able to support services like streaming or IP voice. But still not being able to control the network congestion, this approach can easily degrade the quality and even avoid the access to these services. Emerging services with more demanding QoS requirements, and the need to have control over the performance of these services, are nowadays creating the need for a better QoS management in the network.

From the fixed network point of view, increasing the backbone capacity and user access bandwidth is a necessary first step for accommodating these real-time applications, but it is not the final solution since lower-priority traffic may still flood the network, utilizing all the available resources. The real challenge is to manage the existing network bandwidth, so that a QoS can be delivered to all users.

In wireless networks, radio resources are usually more scarce and expensive than in fixed networks, thus an appropriate resource management is essential. While the circuit-switched approach is based on a permanent and exclusive use of the radio channel as long as the connection is active, packet-switched networks allow several data flows to be multiplexed, sharing the same resources along the time. On one hand, packet-switched approach optimizes the

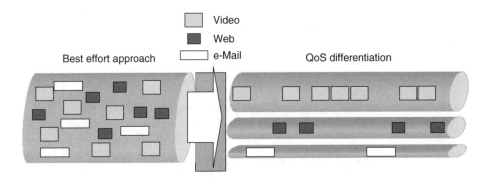

Figure 4.1   QoS differentiation

resources usage due to the *statistical multiplexing gain*,[1] but it requires advanced techniques to provide the right resources to the different data services so that the application is able to perform as expected.

The first step to define a QoS technique consist in the creation of a set of parameters associated to each data flow, which must be specified and negotiated during an initial phase of the transmission. Delay, bandwidth and reliability attributes are commonly used for that purpose. This is especially important for supporting the new generation of Internet and mobile applications such as video streaming and other real-time services.

The final goal of the QoS is to be able to classify the packets, and treat them according to the application requirements so that the maximum number of users is satisfied.

Figure 4.1 represents the classification done by a QoS policy which gives more preference to the packets of more demanding service (video) in order to guarantee a minimum delay, while less bandwidth is given to e-mail service, where users' perception of delay is not so critical.

4.1.3 QoS Standardization

As wireless data services makes use of both the cellular networks and the IP-based Internet for providing the service, both set of protocols and functionalities must cooperate to provide a seamless end-to-end QoS. Both IETF and 3GPP/3GPP2 are contributing to the standardization of QoS-oriented mechanisms and protocols in IP and cellular networks.

Figure 4.2 represents the scope of 3GPP and IETF in the end-to-end path of a mobile network. It can be observed how 3GPP is initially covering the whole wireless network, towards the interfaces to Internet. However, as IP-based Radio Access Networks (RANs) will be developed in the near future, IETF mechanisms and procedures will become more relevant in the wireless side, and the coordination between both systems will be critical.

---

[1] The term 'statistical multiplexing gain' is commonly used to represent the network capacity gain when several independent data sources are sharing a channel instead of using dedicated channels due to the burstiness of traffic sources.

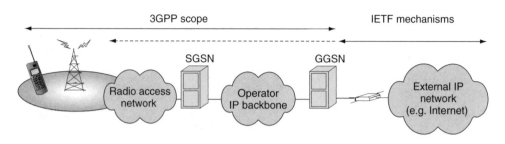

Figure 4.2   End-to-end QoS management (3GPP example)

In the IP-world, many different QoS solutions are recommended in the IETF, where Differentiated Services (DiffServ), Integrated Services (IntServ) and Multi-Protocol Label Switching (MPLS) are the most relevant strategies (section 4.2). Some of these solutions, as well as other type of quality control mechanisms, are described in Internet RFCs and are also being applied within mobile networks. These standardized mechanisms are focused on the interoperability purpose with external networks and QoS management in IP-based intranets or other network elements (e.g. queuing management methods).

In the mobile area, the first QoS framework was defined in 3GPP Release 97, which included basic procedures for establishing logical channels with negotiated QoS along the mobile network. These mechanisms are further described in section 4.3.

Since a huge diversity of QoS mechanisms are simultaneously involved along the end-to-end path, it is crucial to align and map the QoS requirements through the different domains and layers. This is achieved by means of SLAs between domains, which ensure that the end-to-end QoS is properly budgeted along the network.

It is important to note that each solution may use a particular set of QoS parameters at different points of the network. This can cause problems in some cases, as the translation of service requirements into domains is not so straightforward.

The gateway interconnecting the mobile network with the external IP network (e.g. Gateway GPRS Support Node, GGSN, in 3GPP architecture) is responsible for the mapping between the cellular QoS parameters and the IP QoS parameters, and plays a very important role in maintaining a consistent service level between both domains.

4.1.4 Data Services Classification

A proper classification of data services according to their QoS requirements is very important so that the network may treat them accordingly, sharing adequately the available resources among the different data flows. The final objective is to increase the number of satisfied users, according to a given criteria established by the network operator. These criteria can be based on different business strategies, and not necessarily imply that maximum number of users are satisfied. For instance, the objective might be to maximize the number of satisfied users of a certain demanding service (i.e. video streaming), while not degrading other less-demanding services too much (i.e. e-mail delay). The QoS parameters would allow controlling the weight and quality criteria for each of the services.

Table 4.1 Example of data services classification

Service	Reliability	Delay	Guaranteed bit rate
e-Mail	High	High	No
Fax	Low	High	No
Web/WAP browsing	High	Medium	No
Audio/video streaming	Low	Medium	Yes
Audio/video conference	Low	Low	Yes

A basic classification of packet data services can be based on the following requirements.

- *Reliability*: It is related to the importance of packet loss. Some applications are relatively error-tolerant without degrading the end-user perception (e.g. voice audio service). However, other services do not admit any information error or loss (e.g. it would not be acceptable for an end-user to receive an incomplete or erroneous e-mail).
- *Urgency*: It is related to the maximum transmission delay between peers. Applications in where human beings interact (e.g. videoconferencing) would not be feasible in networks with very long delays due to the strict interactivity nature of this service. On the other side, streaming of a radio station, e-mail or file downloading services do not require such interactivity, as the user normally expects certain delay in the service.

Usually, services requiring high urgency but loose reliability characteristics are called real-time services. Examples of real-time services are: audio/video streaming, voice over IP (VoIP), videoconference or Push-to-Talk over Cellular (PoC). Typically, this type of services also requires a guaranteed bit rate, determined by the audio and/or video codec (e.g. AMR for voice, H.263 for video, etc.).

On the other hand, non-real-time services require a high reliability although a low urgency is acceptable. Examples of non-real-time services are: Web or WAP browsing, chat, multimedia messaging or file downloads. Note that some of the aforementioned services (e.g. Web, WAP, chat) have a request–response pattern, so the network should prioritize them against non-interactive services.

Some examples of data service requirements are shown in Table 4.1.

The remaining chapter is organized as follows. Section 4.2 describes the QoS mechanisms in IP-based networks. Section 4.3 outlines the QoS standardization activities within 3GPP and 3GPP2 and provides some guidelines on how to provide QoS differentiation over cellular networks. Finally, section 4.4 describes the QoS policy management architecture and possible implementation in cellular environments.

4.2 IP-Based QoS

Multitude of data applications running on top of IP networks are arising over the Internet, and therefore the mobile networks must include all the infrastructure and transmission capabilities needed to ensure an acceptable service performance. Wireless access to the Internet, either in reduced Wireless Fidelity environments or through cellular networks, is just the ultimate and most flexible way to access to the IP world, and therefore suffers the same requirements than any fixed network.

Each application may generate totally different traffic characteristics, which according to the classification made in section 4.1.4, will have a different set of QoS requirements. Legacy networks need to be adapted to support these new requirements by the introduction of specific mechanisms responsible for assuring that those QoS requirements are fulfilled. Of course, QoS policies cannot solve all the problems added by the more demanding capacity of real-time services, and increasing the network bandwidth and switching capacity will always be needed when traffic load growth explodes. However, what can be done is to maximize the user experience, in terms of satisfied users, with the available capacity.

This section describes the fundamentals of the most important QoS mechanisms and paradigms that are actually used in fixed networks, such as the Internet, and how they can be applied in cellular networks. Benefits and drawbacks of the different alternatives, as well as the network elements that will be affected by those techniques will be described.

4.2.1 Motivation of IP QoS Mechanisms

Internet Protocol (IP) has become a standard for the transmission of almost all data services, including not only non-real-time traffic as for example FTP or HTTP, but also new real-time traffic services, such as VoIP or streaming.

Traditional IP routers do not provide any QoS differentiation among the different packets they handle and all packets are being considered with the same priority, i.e. all traffic is considered as BE. With this approach, it is not possible to provide any type of guarantee in terms of neither packet loss, delay, nor throughput. The main reason for this BE treatment is that it helps to keep the complexity of the communication in end hosts, so the network can stay relatively simple. However, the consequence is that, when more hosts are connected to the network, the services are not denied, but their quality is degraded. It will be a decision of the end peers to decide where the received quality is so poor that the service shall be dropped or denied.

This approach has several limitations, but one of the most important is that delivery delays and variation of delays can be very prejudicial for those applications with real-time requirements, such as telephony or streaming services. A short-term solution for real-time applications can rely on an increase of transmission bandwidth, although with that approach it is still not possible to avoid packet delay jitter during the session.

The BE strategy provides an acceptable performance for the type or services supported in the initial phases of the Internet. However, BE is not enough to support real-time services, and hence the introduction of IP QoS mechanisms is needed, which will provide some level of predictability and control over the available resources. This is especially important over wireless networks, where over-dimensioning the radio interface and infrastructure are much more costly than in fixed networks.

From the fixed network point of view, the easiest way to provide a certain QoS is to ensure the maximum bandwidth is available at all times to all users, but this is not possible, as each network element has a capacity limitation. Additional bandwidth can always be added (if the construction cost is not considered), but as fast as extra bandwidth is added, it is likely to be used sooner or later. Hence, the real challenge is to manage the existing network bandwidth in a suitable manner, which allows that the QoS requirements of a variety of services are fulfilled.

Among the different QoS techniques (also known as QoS paradigms) that can be found in bibliography for QoS differentiation, IntServ, and DiffServ are two of the most well known, but all of them have in common the basic premise of leaving the complexity of the algorithms at the edge of the networks and try to maintain core network as simplest as possible.

4.2.2 QoS Paradigms

4.2.2.1 Integrated Services

IntServ [1] is a QoS paradigm of RR, in which network resources are apportioned according to an application QoS' request, and subject to a bandwidth management policy. The idea behind IntServ is to merge the advantages of both datagram and circuit-switched networks: datagram networks maximize network utilization and provide multi-point communication and robustness by adapting to network dynamics, but do not guarantee a certain service level. On the other hand, circuit-switched networks can provide guaranteed service level by reserving resources for one communication, but may not use them efficiently. IntServ defines three different service models.

1. *Guaranteed service [2]*: It provides an assured bandwidth that produces a delay-bounded service with no queuing losses for all conforming packets of a flow. Guaranteed service is not in charge of controlling minimum or average transfer delay, but only maximum queuing delay, hence jitter (difference between minimum and maximum delay) is neither controlled. In order to provide a guaranteed service, the application shall provide the network with an estimation of the traffic that is going to generate (Tspec parameters), as well as with information of the desired service (Rspec parameters). Tspec and Rspec provide information such as peak rates, minimum policed unit and maximum datagram size [2].
2. *Controlled load [3]*: This model approximates BE scheme under unloaded networks. It is better than BE service, but it cannot provide the strict bounded service that guaranteed service provides. Assuming that the network is working properly, it is possible to consider that applications defined as controlled-load service model will fulfil two common premises.

   (i) Most of the transmitted packets will be successfully delivered by the network to the end nodes.
   (ii) The delay experienced by a very high percentage of the delivered packets will not greatly exceed the minimum delay experienced by an unsuccessfully delivered packet.

In a similar way as in guaranteed service, applications requesting for a controlled-load service level should provide the network with an estimation of the data traffic that they are going to generate via Tspec parameters. The return is that the service level ensures that the network will have the available needed elements in order to provide with such traffic characteristics. If the traffic generated by the applications exceeds the estimations provided by Tspec parameters, the QoS provided by the network could be degraded from the expectations characteristic (e.g. higher percentage of packet loss or high transmission delay).

3. *BE*: BE model, as described in previous sections, do not include any kind of quality control notification or negotiation, and each service will perform correctly as far as the supported load and congestion is kept under certain limits.

In order to make the reservations for a unicast or multicast flow along a determined path in IntServ, the IETF defined a signalling protocol called Resource reSerVation Protocol (RSVP) [4–6]. RSVP is a control protocol that enables user applications to request from the network specific QoS for each data flow. With RSVP, sender and receiver are separated by unidirectional and logically independent flows. In the seven layers of the OSI protocol model, RSVP is located in the transport layer, in the interface with the network layer routing protocols. Generally, RSVP can be seen as a signalling protocol that does not carry user data.

RSVP is in charge of reserving the needed resources, such as bandwidth on an interface, in every intermediate router along the application's data path. This reservation is made during session set-up, and if it is successful, the user will receive a dedicated stream of guaranteed fixed bandwidth for the duration of the session. Once the session is finished, the reserved resources will be returned to the system.

QoS specific paths are defined by sending 'PATH' messages from the original host to the destination in a hop-by-hop fashion. Upon receiving these PATH messages, the receiver will have to acknowledge QoS reservation at each hop to the sender. Receiver acknowledgements are performed via the RESV messages in the reverse direction to the data flow (Figure 4.3).

RSVP alone is not able to provide a better QoS to a certain user, as it is only a control protocol which sets up a reservation of transmission resources. Enforcement of such reservation needs to be performed by some other components of the network architecture. This is typically done by making use of routers with sophisticated scheduling techniques, which are able to provide a service level according to the reservations previously done.

Reservation procedures, management and enforcement of bandwidth consume processing power in the different routers. In addition, each router along the patch needs to reserve information concerning the flows to be transmitted through them, as traffic description or address of next RSVP host in the path. The amount of this information is proportional to the number of reservations, and in case a router handles tens of thousands of flows, keeping state information can become a problem. On large interfaces, the bottleneck is usually processing speed of the routers, rather than the bandwidth of the outgoing link.

Functional characteristic of reserving certain resources for a connection make this solution suitable for those services requiring a certain guaranteed QoS requirement. Typical services that would benefit from RSVP are conversational (like VoIP or videoconference) and streaming services (see section 4.3 for further details on services classification). However,

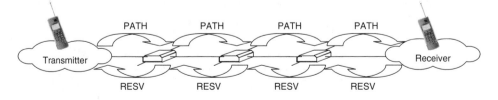

Figure 4.3   RSVP path creation

these characteristics make RSVP not to perform very well under cellular networks due to the reasons listed below.

- In order to maintain RSVP path along a session, the sender needs to refresh the state of the reserved link by sending PATH messages to the receiver every 30 seconds. This constant flood of messages between the different RSVP routers will increase load to the underlying transport network.
- It does not provide fast handovers that are required in wireless networks. A handover entails a change of the base station the mobile is connected to, and consequently, a change in the whole path from the mobile to the other end. If any of the IP-based network elements along the actual RSVP path is changed due to the handover, this change of path leads to a release of the resources along the actual RSVP path and a new RSVP path is established through the new base station. This process of releasing and establishing a new path takes some time, which can be too high for some real-time services, as this delay needs to be summed up to the delay of the handover procedure itself.

4.2.2.2 Differentiated Services

DiffServ [7,8] is a QoS paradigm that provides a method for classifying services of several applications. This is done by providing different levels of service to different packets, by means of the DiffServ CodePoint (DSCP). DSCP are included as part of the header of each IP packet. In IPv4 the information is stored in the TOS field, while in IPv6 header the Traffic Class octet is used.

The DiffServ QoS mechanisms are applied in edge routers, and consist of (Figure 4.4):

- *Packet classifier* – in charge of identifying the different incoming flows to the cellular core network.
- *Traffic marker* – in charge of introducing DSCP in all incoming IP packets according to assigned class.
- *Traffic meter* – in charge of monitoring traffic flow of a determined user.
- *Traffic shaper* – in charge of reacting for traffic exceeding QoS requirements of a certain user.

More details on aforementioned QoS mechanisms are explained in section 4.2.4.

Inside a DiffServ network, all packets marked with the same service class have to get the same treatment by all the routers within the network. That is called Per-Hop-Behaviour (PHB). PHB provides three different service levels.

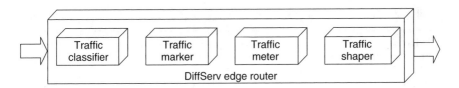

Figure 4.4 DiffServ edge router block diagram

1. *Expedited forwarding (EF)* [9]: EF is very similar to Constant Bit Rate service in an Asynchronous Transfer Mode (ATM) network, and its idea is trying to guarantee the delivery of the packets, minimizing delay and jitter, and with a very small packet loss ratio. Provided that the traffic sent to the network does not exceed a predetermined limit. Traffic exceeding such limit is discarded. EF defines a unique codepoint, and it is typically applied to real-time flows, as VoIP. EF PHB is not mandatory part of DiffServ architecture, i.e. a node is not required to implement EF PHB to be DiffServ compliant.
2. *Assured forwarding (AF)* [10]: AF provides looser guarantees in terms of packet loss and delay, but it allows a more flexible implementation. This is done by providing four classes and three drop-precedences within each class (a total of 12 codepoints). Excess traffic above traffic profile requirements is not discarded, but it is provided with a lower probability than the traffic within traffic profile. Different drop precedences within one class are treated with the same priority under non-congested situations, but those precedences can be used at DiffServ nodes to reduce or even eliminate congestion by discarding packets marked with high drop precedence. AF is typically applied to interactive flows, as Web services.
3. *BE*: No priorities are provided to the packets, but the service performance will directly depend on the network status.

PHB is applied by traffic conditioner at a network input point, according to a predetermined policy criterion. Traffic is marked at this point (DSCP are included in IP header of all IP packets), and internal routing is done according to such marking. When the packet reaches the network border exit, it is unmarked. This means that in DiffServ, routers do not have to store any state information of the flows.

DiffServ can be easily applied to cellular networks by using EF, AF and BE PHB classes. PHB classes can be directly mapped onto UMTS traffic classes. Table 4.2 shows an example of such mapping. In this table, DSCP are defined as recommended by [10] and [9].

In all traffic classes mapped onto AF PHB classes, traffic within traffic profile requirements should be provided with the lowest drop precedence within the corresponding AF Class. Streaming traffic exceeding subscribed values should be assigned with higher drop precedence within Class 1. The same behaviour is applied to interactive THP1 traffic, but in Class 2. Interactive THP2 and THP3 traffic exceeding subscribed one could be provided with high drop precedence or it could be assigned as BE traffic. In case of conversational class, traffic exceeding subscribed values is discarded.

The main advantage of this traffic mapping is the fact of not introducing extra signalling along the paths (as it is the case of IntServ). The only needed signalling would be between the host and the edge router, but due to the use of UMTS traffic classes in the cellular network, packets coming to edge routers are already assigned to a certain traffic class. Therefore, a simple mapping table in the edge routers is required, as the one shown in Table 4.1.

Table 4.2 Example of UMTS QoS classes mapping to DiffServ classes and code points

UMTS traffic class	DiffServ service level	DSCP
Conversational	EF PHB	101110
Streaming	AF PHB Class 1	001010
Interactive (THP1)	AF PHB Class 2	010010
Interactive (THP2, THP3)	AF PHB Class 3	011010
Background	BE	000000

In order to implement DiffServ in a UMTS network, the operator needs to support both edge and core DiffServ capable routers. Edge and core routers can be physically different network elements or they can refer to a set of control functions, depending on the network topology. Edge routers need to be placed in those network elements that insert IP traffic to the network, as for example GGSN. Core routers need to be defined in those places requiring IP routing function, which typically are routers within the core network and RAN.

The expected main advantage of DiffServ solution is that it provides a scalable implementation, compared to IntServ. Furthermore, it is supposed to provide an efficient multiplexing of services and applications. Main drawback is that standardization process of DiffServ is not totally following those objectives, but proposed solutions are a bit more complex, similar to ATM standards. Additionally, DiffServ has an important drawback compared to IntServ: it has no end-to-end control in terms of performance, but it is based on a PHB approach. Therefore, end-user performance depends quite a lot on the network status and no guarantees can be fulfilled.

4.2.2.3 Multiprotocol Label Switching (MPLS)

MPLS is a forwarding protocol defined by the IETF based on the use of labels to identify streams [11]. The fundamental idea is to assign short fixed-length labels to the different packets at the input of a MPLS network. Inside the MPLS domain, packets are forwarded based on this label, without using original packet headers at all. MPLS marking functionality is designed to determine the next router hop, unlike for example DiffServ, which makes use of marking to determine priority within a router. In addition, label information is also used to identify QoS classes. In MPLS, packets with the same label follow the same path, known as LSP (Label Switching Path).

MPLS is a protocol applied only on routers (Label Switching Router, LSR), in which the forwarding decision is done based on the destination address. The first LSR in the network attaches the label to the packet and forwards it to the next hop. At this next hop, label is used as an index to a table that specifies the next label and next hop to be used. This procedure is done in all routers within the MPLS network until the output router is reached, and the labels are removed. Figure 4.5 represents this procedure.

LSR1				LSR2				LSR3			
In label	Address prefix	Out interface	Out label	In label	Address prefix	Out interface	Out label	In label	Address prefix	Out interface	Out label
—	192.168.	1	9	9	192.168.	1	7	7	192.168.	1	—
—	127.0.	1	8	8	127.0.	0	6	2	127.0.	0	—
—	—	—	—	7	192.168	0	5	—	—	—	—

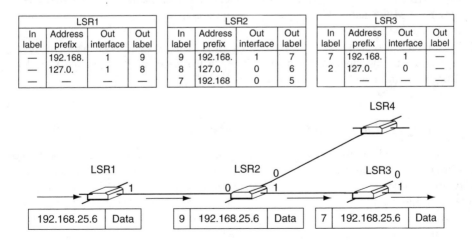

Figure 4.5 MPLS packet forwarding

A mechanism for label distribution, Label Distribution Protocol (LDP) is needed in order to build the routing tables. LDP is in charge of negotiating labels between neighbour routers along a determined LSP, from entrance point to exit point. That way, labels have only local significance. MPLS includes also the possibility of performing explicit routing, in which the edge router at the entrance of the network determines the path to be used up to the destination, indicating which nodes to be used on that route.

4.2.2.4 Combined Solutions

Paradigms showed in previous chapters are typically combined in real networks in order to provide a further control over QoS requirements. Main combinations used are presented in this section: IntServ + DiffServ, MPLS + RSVP, MPLS + DiffServ.

IntServ + DiffServ: The idea of this solution is try to combine the benefits of both alternatives. The basic way of working is that IntServ is used wherever it is possible in the network (providing hard guarantees) and DiffServ is used just in those places in which it is not possible to reserve resources, as for example in the backbone area.

MPLS + RSVP: This combination makes use of MPLS, but doing at the same time a set-up of QoS reservations. Instead of using LDP protocol to establish MPLS paths, those are established via RSVP PATH and RESV messages. Using RSVP, multiple traffic LSPs could be established between two distinct nodes. In addition, RSVP could also be used to establish LSP with or without RR. For BE traffic LSP without RR is adequate. However, LSPs with RR can be used to route real-time traffic.

MPLS + DiffServ: In this combination, MPLS paths (LSP) can be used to map DiffServ PHB aggregates, i.e. one LSP can be defined for each type of service. Like MPLS labels, DSCP are marked at the edge router. DSCP are used within the DiffServ network to implement suitable scheduling algorithms and some time drop precedences. Two different solutions can be applied.

1. Several PHB classes are mapped within the same LSP: In this case, the edge router marks each PHB class with a unique EXP (Experimental) bits within the MPLS header. The LSR within the MPLS network can identify the appropriate PHB class from the EXP bits and treat them accordingly.
2. To establish a separate path for each PHB class: This is established between the edge routers during the label establishment phase. In that case, EXP can be used to identify drop precedences within the same PHB class.

4.2.3 IP-QoS Management in UMTS Networks

QoS provisioning is performed in different domains, and some of them are not controlled by the UMTS system. This creates the needed to inter-work between the UMTS and the external network in order to create a consistent management of the connection resources. This inter-working can be done in different ways, depending on the QoS mechanism applied in the external network. Those networks are typically IP-based networks, and hence, IP-QoS mechanisms explained in this section are applied.

Table 4.3 IP BS Manager capabilities

Capability	UE	GGSN
DiffServ Edge	Optional	Required
RSVP/IntServ	Optional	Optional
IP Policy enforcement point	Optional	Required

A network element called IP BS Manager is defined in Release 5 of UMTS specification in order to manage end-to-end IP QoS. This IP BS Manager uses standard IP mechanisms to control the IP Bearer Services. This entity, which is not mandatory in the network, is included as part of the UE and the GGSN, and it may support both DiffServ Edge function and RSVP. If IP BS Manager is not implemented in the network, QoS provisioning support can be done by other functionalities in the GGSN and UE ends.

IP BS Manager makes use of the translation/mapping function to perform the mapping between parameters used in the UMTS Bearer Service and those used in the IP Bearer Service and vice versa. IP Bearer Service parameters are mapped onto UMTS QoS parameters in the GGSN end, whilst they are obtained from application QoS parameters in the UE end.

The 3GPP standards [12] defined the minimum capability an IP BS Manager node shall have in case it is implemented in a UMTS network. Those capabilities are different in the UE and GGSN, and depending on the features that are implemented in the network, different working scenarios will be applicable. Table 4.3 summarizes the different alternatives.

The DiffServ Edge function is mandatory in the GGSN, although the way it does the mapping onto/from UMTS QoS parameters is a network vendor solution. The only requirement is that it shall be compliant with IETF specifications for DiffServ. Parameters used for the DiffServ function, as those needed in traffic classifier, marker, meter and shaper functionalities maybe statically configured in the GGSN. They can be derived from PDP context parameters, or they can be obtained from RSVP signalling (in case combined DiffServ + IntServ is used). In the UE end, DiffServ function acts as a boundary for the incoming traffic from application layer. One important thing is that when DiffServ marking function is implemented in the GGSN, DiffServ function in UE is not needed.

Regarding RSVP support, functional description in the GGSN is not defined as the standards, but mainly it should be based on the IETF specifications for IntServ. In the UE end, RSVP function provides the UE the possibility to request end-to-end QoS by using RSVP messages defined by IETF specifications.

4.2.4 Traffic Handling Mechanisms

All the QoS paradigms described in the previous chapter need to be implemented by means of traffic handling mechanisms, which are in charge of the treatment of the transmitted packets: reservation, dropping, allocation, etc.

4.2.4.1 Packet Classifier

Packet classifier is the mechanism in charge of mapping incoming packets onto traffic classes that can be internally understood by the rest of traffic handling mechanisms. Those traffic classes are characterized by the fact that all packets belonging to the same traffic class shall be treated with the same priority by the packet scheduler.

Packet classifier is used specially in network edge routers, both input and output ones. This functionality is used, for example, in DiffServ Edge router to map from DSCP to UMTS traffic class and vice versa, using mapping tables as the one described in Table 4.2.

4.2.4.2 Packet Scheduler

Packet scheduler is the functionality in charge of forwarding all incoming packets to a certain element, typically a router, according to a certain priority scheme. Schedulers make use of different queues, typically one per priority, to classify the packets and serve them with the appropriated priority to the destination router or host.

There are a lot of scheduling methods defined in the literature, depending on the quality level that is desired. From very simple algorithms in which all packets are provided with the same priority (e.g. Round Robin) to complex algorithms in which not only a priority is provided, but also throughput or delay guarantees can be provided (e.g. Class-based Queuing [13]).

4.2.4.3 Admission Control (AC)

AC is the functionality in charge of deciding if a new packet flow can be granted with the needed resources according to its QoS requirements or not. Additionally, AC shall consider if the introduction of a new flow might affect the quality of the users that are already active in the system and deny the access to new users, which would overcome the target quality criteria.

In a similar way as with packet scheduler, there are a lot of AC policies, all of them having the final goal of providing the QoS requirements requested by incoming flows. For instance, admission of high priority users should be prioritized against low priority ones, or, as described in [14], AC can be done in a different way for real-time and non-real-time users.

4.2.4.4 Packet Dropping

Packet dropping is the network functionality in charge of discarding packets of certain flows when congestion is reached in the network or when a certain traffic flow is exceeding its assigned resources (e.g. DiffServ traffic with AF PHB class exceeding subscribed bit rate). Dropping of packets will be done according to some dropping precedence, from less important packets to most important ones.

This mechanism is quite important for preventing congestion in buffers, typically produced by the so-called resource eating traffic, as for example TCP traffic. Packet dropping is applied on router buffers, and there are different policies defined in the bibliography. Weighted Random Early Detection (WRED) [15] is a very common solution, as it allows the use of DSCP for calculating the drop probability of a packet.

4.3 QoS Architecture in 3GPP and 3GPP2

4.3.1 End-to-End QoS Introduction

The concept of QoS has been created and defined in order to provide different treatments to traffic from different services and/or users. QoS is a broad concept that can be decomposed in the QoS provided in the different domains and layers of the system, but that has significance only when considered end-to-end, as the final QoS provided is always determined by the weakest layer [16].

More specifically, QoS mechanisms are responsible for the traffic management and service differentiation from the network point of view, i.e. according to certain quality requirements specified by the application, the network functionalities treat each service according to their own requirements. However, the perception of service quality from end-user point of view is a wider and more subjective issue, also defined as Quality of Experience (QoE). In any case, from the cellular network perspective the best way to try to ensure a good QoE is to provide a good end-to-end QoS.

QoS management architecture defined in the different standardization forums such as 3GPP and 3GPP2 have defined the end-to-end QoS as a combination of QoS provisioning in different layers of the subsystem. This common approach is shown in Figure 4.6.

As illustrated in Figure 4.7, end-to-end QoS is achieved by mapping the end-to-end requirements onto a proper configuration along the different mechanisms available in the network. For instance, the RAN includes QoS-aware mechanisms like AC, channel allocation, packet scheduler, power control or handover control that may differentiate among services and/or users. In the core network, there exist many other mechanisms such as congestion control, authorization of resources, RR for real-time services, etc. Additionally, the operator

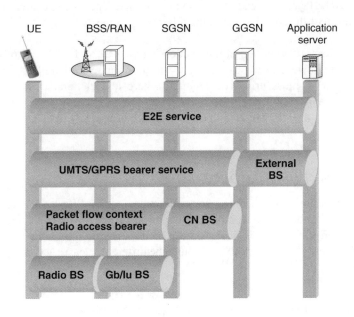

Figure 4.6   End-to-end QoS architecture

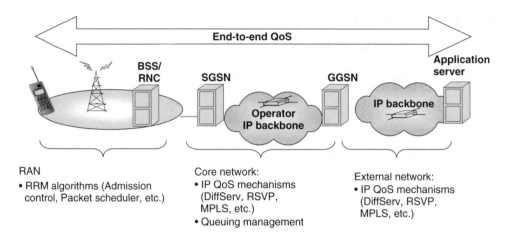

Figure 4.7   Main QoS Mechanisms to provide end-to-end QoS

may include any IP-based QoS method within its own intranet as well as a proper mapping between mobile and external networks.[2]

The end-to-end service is defined on top of UMTS/GPRS bearers and external bearers. The UMTS/GPRS bearers follow a bottom-up philosophy, being the upper bearers defined on top of the capacity provided by the lower bearers. As an example, UMTS bearer is compound of radio access bearer and core network bearer; while the radio access bearer is compound of a radio bearer and Iu bearer.

Each of the bearers shown on Figure 4.6 have their own QoS mechanism; therefore QoS provisioning can be layered or classified also by QoS mechanisms per bearer, being end-to-end QoS the result of the combination of the different QoS mechanisms through the different system bearers.

The QoS architecture described in this section is now part of the Release 6 of 3GPP and 3GPP2 standards. However, this structure is the result of an evolution of the specifications, which comes from a more primitive solution, where less elaborated mechanisms were implemented and only few bearers were available. Currently, most legacy live 2 and 2.5 Generation networks are still working on a BE approach or use very basic QoS differentiation based on priorities.

4.3.2 Evolution of QoS in 3GPP Releases

3GPP specification aimed at providing a common framework for different telecommunications standards. These standards are in constant evolution driven by market requirements and technologies improvements. This evolution can be also seen in the QoS architecture of the 3GPP specifications, which has evolved through the different releases according to the needs of new services [17].

---

[2] Note that although External Bearer Service is not under the scope of the 3GPP specification, it includes the requirements from the mobile network elements to provide a translation from mobile to IP-based external networks [12].

4.3.2.1 Release 97/98

The introduction of GPRS Bearer Service in Rel'97 meant the first step in the evolution of the QoS management architecture for mobile networks. The new bearer service provided by GPRS was based on the concept of PDP context, which is, in essence, a logical connection set-up between a Mobile Station (MS) and a GGSN to carry all the IP traffic generated to and from that MS. The QoS architecture in Rel' 97/98 is depicted in Figure 4.8.

The QoS architecture relies on the PDP context concept to provide the differentiation between data packets. The QoS parameters are established on a PDP-context basis, so all the traffic that is sent via the same PDP context has the same QoS, even if that traffic comes from different applications launched by the same user. In order to achieve a differentiated QoS treatment for the same MS, several PDP contexts can be activated, so that each one can have its own QoS profile. However, in this release the terminal receives a new PDP address per PDP context, which can be a clear limiting factor for services like streaming, where several PDP contexts with different QoS profiles must be active simultaneously for the same application to carry interactive and real-time traffic.

The QoS attributes are negotiated between the terminal and the network during the PDP context activation procedure. In Rel'97/98 the QoS profile is defined according to the following attributes: *precedence, delay, reliability* and *throughput classes* (Table 4.4). The terminal is the one selecting the values of the QoS attributes according to the application needs. In case any of the values is not specified by the terminal, the network will be responsible for adding a default value in the QoS profile.

In Rel'97/98, only non-real-time services are supported, such as e-mail, Web browsing, file transfer, Multimedia Messaging Service (MMS) or Wireless Application Protocol (WAP) services.

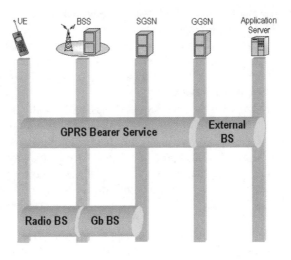

Figure 4.8   QoS architecture in Rel'97/98

Table 4.4   QoS attributes in Rel'97/98

QoS attribute	Description
Precedence class	There are three different service precedence levels, which indicate the priority of maintaining the service. High precedence level guarantees service ahead of all other precedence levels. Normal precedence level guarantees service ahead of low priority users. Low precedence level receives service after the high and normal priority commitments have been fulfilled.
Delay class	Delay attribute refers to end-to-end transfer delay through the GPRS system. There are three predictive delay classes and one BE class.
Reliability class	There are five different reliability classes. Data reliability is defined in terms of residual error rates for the following cases: probability of data loss, probability of data delivered out of sequence, probability of duplicate data delivery and probability of corrupted data.
Throughput class	User data throughput is specified in terms of a set of throughput classes and characterizes the expected bandwidth required for a PDP context. The throughput is defined by both peak and mean classes. The peak throughput specifies the maximum rate at which data are expected to be transferred across the network for an individual PDP context. There is no guarantee that this peak rate can be achieved or sustained for any time period; this depends upon the MS capabilities and available radio resources. The peak throughput is independent of the delay class that determines the per-packet GPRS network transit delay. The mean throughput specifies the average rate at which data are expected to be transferred across the GPRS network during the remaining lifetime of an activated PDP context. A BE mean throughput class may be negotiated, and means that a throughput shall be made available to the MS on a per need and availability basis.

4.3.2.2 Release 99

Rel'99 enables the usage of several PDP contexts per PDP address, each having a QoS profile of its own. The first PDP context activated for a PDP address is called *primary* or *default PDP context*. The subsequent PDP contexts activated for the same PDP address are called *secondary PDP contexts* as long as they are connected to the same Access Point Name (APN) as the primary PDP context. Otherwise, a new primary PDP context is activated for this terminal pointing to a different APN.

The capability of establishing several simultaneous PDP contexts with different QoS characteristics using just one IP address is feasible thanks to the introduction of the Traffic Flow Template (TFT) [18]. The TFT is based on different combinations of filters like source address and subnet mask, destination and/or source port range, etc., which are defined by the terminal at the time of activating the PDP context. A TFT is assigned to all secondary PDP contexts in order to direct the incoming downlink traffic to the right PDP context in the GGSN, being the primary PDP used as default one (in case the incoming traffic is not matching with any of the existing filters).

In addition, Rel'99 meant to harmonize the GERAN and UTRAN QoS management frameworks, i.e. to align the GPRS bearer with the one defined for UMTS and provide the same QoS parameters and procedures. Rel'99 introduced new attributes, which are described in Table 4.5 and the UMTS QoS classes: *Conversational* class, *Streaming* class, *Interactive*

Table 4.5   QoS attributes in Rel'99

QoS attribute	Description
Traffic class	Type of application for which the Radio Access Bearer (RAB) service is optimized ('conversational', 'streaming', 'interactive' or 'background').
Delivery order	Indicates whether the bearer shall provided in-sequence Service Data Unit (SDU) delivery or not.
Maximum SDU size (octets)	Defines the maximum allowed SDU size.
SDU format information (bits)	List of possible exact sizes of SDUs. If unequal error protection shall be used by a RAB service, SDU format information defines the exact subflow format of the SDU payload.
Delivery of erroneous SDUs	Indicates whether SDUs with detected errors shall be delivered or not. In case of unequal error protection, the attribute is set per subflow.
Residual Bit Error Ratio (BER)	Specifies the undetected BER for each subflow in the delivered SDUs. For equal error protection, only one value is needed. If no error detection is requested for a subflow, residual BER indicates the BER in that subflow of the delivered SDUs.
SDU error ratio	Defines the fraction of SDUs lost or detected as erroneous. SDU error ratio is defined only for conforming traffic. In case of unequal error protection, SDU error ratio is set per subflow and represents the error ratio in each subflow. SDU error ratio is set only for subflows for which error detection is requested.
Transfer delay (ms)	Indicates the maximum delay for 95th percentile of the distribution of delay for all delivered SDUs during the lifetime or a bearer service, where delay for an SDU is defined as the time from a request to transfer an SDU at one Service Access Point (SAP) to its delivery at the other SAP.
Maximum bit rate (kbps)	Maximum number of bits delivered at a SAP within a period of time divided by the duration of the period.
Guaranteed bit rate (kbps)	Guaranteed number of bits delivered at a SAP within a period of time (provided that there are data to deliver), divided by the duration of the period.
Traffic handling priority	Specifies the relative importance for handling of all SDUs belonging to the RAB compared with the SDUs of other bearers.
Allocation/retention priority	Specifies the relative importance compared with other RABs for allocation and retention of the RAB. The allocation/retention priority attribute is a subscription parameter, which is not negotiated from the mobile terminal.

class and *Background* class [17]. The main distinguishing factor between these classes was how much the traffic is affected by delay and packet loss. Conversational class was meant for traffic which is very delay sensitive while background class was meant for the most delay insensitive traffic.

Conversational and *Streaming* classes [17] were mainly intended to carry real-time traffic flows. Conversational real-time services, like VoIP, videoconference over IP or real-time network gaming, are the most delay-sensitive applications, and those data streams should be carried in conversational class. Streaming class is also meant for delay-sensitive application, audio or video streaming, but unlike conversational applications, they can compensate some delay variation with client-embedded buffering mechanisms. The main difference with conversational applications is the unidirectional nature of streaming, which makes it less interactive.

Interactive and *Background* classes [17] are mainly meant to be used by applications which do not have real-time delay restrictions although they are not error tolerant, like Web or WAP browsing, MMS or e-mail. Owing to looser delay requirements, compared to conversational class, both provide better error rate by means of retransmissions schemes. The main difference between interactive and background classes is that interactive class is aimed to give better round-trip time to services which requires a request/response pattern between the two parties, such as Web or WAP browsing. Background class is used for services which have very loose delay requirements and can be satisfied just with a very basic BE scheme.

The main characteristics and possible applications of the different traffic classes are shown in Figure 4.9.

Apart from the traffic class, there exist other QoS attributes mainly related to delay, bit rate and reliability requirements, which are described in Table 4.5.

Rel'99 also includes some features to allow a better control of the QoS provision in the radio domain. QoS requirements can be sent from the 2G-SGSN to the Base Station Subsystem (BSS) by means of the BSS Packet Flow Context (PFC). A BSS PFC may be shared by several activated PDP contexts with similar QoS requirements. The data transfer related to PDP contexts that share the same PFC constitute one aggregated packet flow, whose QoS requirements are called aggregate BSS QoS profile. The aggregate BSS QoS profile defines the QoS that must be provided by the BSS for a given packet flow between the MS and the SGSN, i.e. for the Um and Gb interfaces combined.

The QoS architecture in Rel'99 is depicted in Figure 4.10.

The UMTS QoS profile of a connection is the result of a QoS negotiation between the UMTS QoS requested by the UE and the one granted by the network. The UMTS QoS negotiation occurs typically at PDP Context Establishment, though it might be renegotiated by means of PDP Context Modification procedure [18].

	Traffic classes	Main characteristics	Possible applications
Real time	Conversational	• Stringent and low delay • Minimize delay variation	• Voice call • Videoconference call
	Streaming	• Minimize delay variation	• Audio streaming • Video streaming
Non-real-time	Interactive	• Request response pattern • Minimize bit error rate	• Web browsing • WAP browsing
	Background	• Destination is not expecting the data within a certain time • Minimize bit error rate	• Mail • File download

Figure 4.9   3GPP UMTS QoS classes

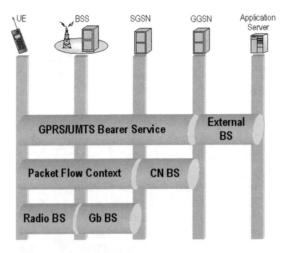

Figure 4.10   QoS architecture in Rel'99

During PDP Context activation procedure, the UE requests to the CN the access to an IP external network, identified by an APN, and includes a UMTS QoS profile request for the PDP context. Before CN answers positively, it needs to perform certain checks.

- Validate the requested QoS profile for the APN against the user subscription in the Home Location Register (HLR). In HLR, the user subscription information is stored in associations where for each authorized APN, the maximum QoS profile that can be granted is indicated. If the requested QoS profile is higher[3] than the one stored in the HLR, then the requested QoS cannot be granted as such, i.e. PDP context may be rejected or QoS may be downgraded.
- Perform AC and RR in CN.
- Trigger lower bearers establishment (e.g. UMTS QoS profile is mapped onto RAB QoS Profile) and QoS checking.

After positive result from lower bearers (RAB and CN Bearer) is obtained, the PDP context is successfully established. However, it may be possible that the requested UMTS QoS and granted UMTS QoS are not the same, due to the limitations introduced by AC in the different network elements and lower bearers.

It is important to note that a PDP context has associated a specific QoS profile, but it can carry data from two or more different applications. Therefore these applications will experience the same QoS, as illustrated in Figure 4.11a. Ideally, for two applications that have very different QoS requirements, it is recommended to use different PDP contexts, even if they are connected to the same APN. With this approach, different QoS profiles per PDP context can

---

[3] A QoS profile is higher to the one in the HLR if it belongs to a higher traffic class (i.e. order is conversational, streaming, interactive and background) or if belonging to the same traffic class, some of the QoS profile parameter is higher.

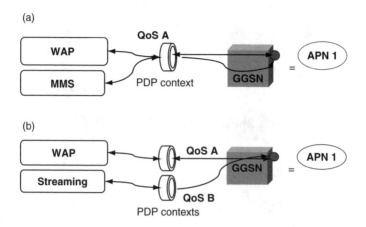

Figure 4.11   Relationship between PDP context, APN and QoS profile. (a) One PDP context for APN 1 with QoS A for two different applications. (b) Two PDP contexts for APN 1 with different QoS profiles for two different applications

be applied. This is represented in Figure 4.11b. However, it is not possible in the scenario where several applications are sharing the same PDP context but pointing to different APNs.

Not all the terminals are able to establish multiple PDP contexts, because it imposes strong complexity in the terminal side. That is the reason why there are research efforts to enable the utilization of a unique APN for different applications, but still providing QoS differentiation. All those studies go towards the installation of functionalities in (or collocated to) the GGSN to analyse the user plane traffic and treat it differently (higher or lower priority) depending on the source application.

4.3.2.3   Release 5

Rel'5 allows negotiating the QoS requirements at application layer, from an end-to-end perspective. With Rel'5 the IP Multimedia Subsystem (IMS) services were introduced and bearer QoS authorization was standardized for them [12]. This authorization provided a new way of activating a secondary PDP context with specific QoS attributes by means of an inter-action between the MS and the GGSN. IMS is explained in detail in section 4.3.3.

The QoS architecture in Rel'5 is depicted in Figure 4.12. An important issue in the overall QoS negotiation (which is also applicable to previous releases) is to perform a proper QoS mapping from high to low bearer services. This means that high-level QoS requirements negotiated at application layer (E2E Service) must be properly mapped onto the PDP context QoS profile as well as external bearer service attributes. Likewise, QoS profile must be mapped onto RAB QoS attributes (or PFC for GPRS) and CN bearer attributes.

The RAB in UMTS or its equivalent in GPRS, PFC, allows a better control of the QoS in the RAN. The QoS attributes for a RAB are common, but the values that are requested shall be adapted to the available radio technology capabilities.

The UMTS QoS attributes, seen as overall level attributes, are usually keeping the same value when mapped onto RAB QoS attributes, however, there are three of them that are usually

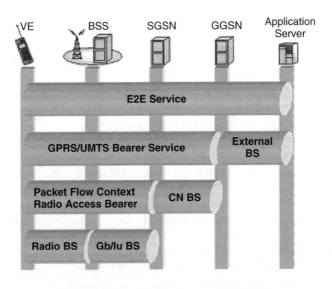

Figure 4.12   QoS architecture in Rel'5

reduced: residual BER, SDU error ratio and transfer delay. The reason for that is that CN Bearer Service may introduce errors and delay, which should not be accounted in the RAB level.

For example, let us consider the case where UMTS QoS attributed 'transfer delay' is 350 ms. This delay may be divided into 250-ms delay requirement for the RAN side, and 100 ms for the CN side. Thus, this would translate to RAB QoS 'transfer delay' of 250 ms (this mapping is implementation dependent).

The CN bearer is established between SGSN and GGSN, typically based on IP or ATM transport. For the IP-based backbone, it is mandatory to use DiffServ as defined in section 4.2. The mappings between UMTS QoS classes and DSCPs are controlled by operator decision. Table 4.2 shows an example on how operators can map UMTS QoS class to DSCPs for the CN Bearer Service, for example it shows how conversational traffic would be DiffServ-marked to be EF (Expedited Forward), i.e. to have the highest priority in the IP backbone.

4.3.2.4   Release 6

In Rel'6, the P-CSCF, which is an architectural element of IMS, becomes the Application Function (AF). Similarly to the P-CSCF, the AF offers services that require the control of IP bearer resources, maps QoS-related application level parameters (e.g. Session Description Profile, SDP) onto policy set-up information, and sends this information to the Policy Decision Function (PDF). In the particular case of IMS, the AF is the P-CSCF which is in the same domain as the GGSN. A new standardized interface Gq, specified in 3GPP TS 23.002, is used for service-based policy set-up information exchange between the PDF and the AF. This information is used by the PDF for service-based local policy decisions.

The main difference between the AF and the P-CSCF is that the AF may be either in the same domain as the PDF or in a different domain, whereas the P-CSCF is always in the same domain as the PDF, which is in turn always in the same domain as the GGSN.

4.3.3 IP Multimedia Subsystem (IMS)

The IMS is an extension of the packet-switched CN, intended to provide a flexible IP media management and session control where operators can expand their new services. IMS is standardized from Rel'5 on, and uses the Session Initiation Protocol (SIP) to set up, maintain and terminate packet-switched voice and multimedia sessions.

SIP is an IETF-driven text-based signalling protocol used for establishing sessions in an IP network, and is also the official call control protocol in 3G networks [19]. It is based on the request–response scheme similar to HTTP signalling. A session could be a simple two-way telephone call or it could be a collaborative multimedia conference session. A single SIP client in the UE is able to support wide range of service capabilities (e.g. VoIP, multimedia, IM, presence, etc.).

It is important to highlight that before Rel'5 the mobile network was totally transparent to the application level negotiation between terminal and server. Somehow, operators wanted to be aware, control, authorize and even charge based on what it is negotiated at that level. In order to do that, new network elements have been standardized to take the responsibility of these functionalities. Another driver for the introduction of the IMS is the possibility that the development of new services provided by the operators is done independent from the available bearer.

The architecture of IMS is made of several network elements that are having, among other functionalities, policy control and QoS functionalities, as shown in Figure 4.13.

When an IMS session is to be established, the UE–AF communication is performed via SIP negotiation, which will contain the set of QoS parameters, including a SDP [20]. In general, SDP contains a description of the media to be transported as well as other type of information like type of codec, port numbers, etc. Based on the SIP negotiation (including SDP information) made between parties, the operator is able to authorize or not the request resources based on a certain policy criteria. Therefore, QoS authorization is performed in the set of common codecs which is supported by all the parties in the communication. The typical SDP description

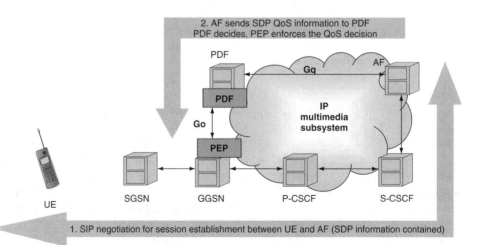

Figure 4.13 IMS architecture

consists of a *session-level description*, containing the common details that apply the whole session and several *media-level descriptions*, which apply to a single media flow. The main components used for mapping a SDP description onto a QoS authorization are Media announcement ('m = '), Connection data ('c = '), Attributes ('a = ') and Bandwidth ('b = ').

If the negotiation is successful, a 'Service-based Local Policy' is established in the IMS, and the session receives the name of '*SBLP*'. This means that the service-level would impose its policies and QoS requirements over the communication. If the service-level does not impose any QoS requirements, then this type of IMS session is known as 'non-SBLP' session. Only in IMS 'SBLP' sessions, end-to-end QoS is provided as QoS is ensured not only in the cellular network but also in the IMS.

4.3.3.1 SBLP Sessions for End-to-End QoS

Only 'SBLP' sessions provide end-to-end QoS, by means of different QoS rules and policies that are located in GGSN, PDF and AF. After the SIP negotiation, AF is the network element providing the QoS requirements and from the application perspective AF indicates the session QoS requirements (i.e. SDP) to the PDF over the Gq interface.

PDF is the network element that authorizes AF QoS requirements, by the creation of IP QoS rules to be applied on the user plane. These rules would set limits to the authorized overall session bandwidth at IP level, the individual bandwidth of each IP flow, the IP addresses that can be accessed, etc.

GGSN interface towards the IMS side has a PEP (Policy Enforcement Point, see also section 4.4) and DiffServ functionality (RSVP is optional). The PEP functionality is only used for 'SLBP Sessions', and represents a 'gate' in the GGSN that is open or closed depending on the success of the session QoS authorization performed at the PDF.

When the QoS authorization is successful, PDF over 'Go' interface notifies to the GGSN the IP QoS rules and authorized QoS. Subsequently, GGSN 'opens' the gate for the authorized flow. On the other hand, GGSN has a translation/mapping function that translates the authorized QoS to UMTS QoS. A detailed description of this phase is shown in the flowchart of Figure 4.14 [21], including the following procedures.

1. A primary PDP context is used to convey the SIP signalling, which is established after the terminal is attached to the 3G mobile network and before it performs the CSCF discovery procedure.
2. A SIP dialogue between the UE and the server, intermediated by the CSCF, occurs on the application level to establish the SIP session. In this step, the negotiation of application parameters that have an effect on the QoS requirements is performed (e.g. media encoding).
3. A secondary PDP context activation is then requested with the QoS requirements negotiated during the SIP signalling. In this case, resource reservations are necessary to support real-time requirements. Therefore, the key issue in the secondary PDP context establishment is how the AC and RR are performed at each network element.
4. The RAN performs an AC basically based on the availability of radio resources.
5. Once the local AC is performed in the GGSN, it outsources another AC coordinated with the PCF in the P-CSCF by submitting a policy request to the PCF by sending a COPS message. Once the resources are authorized, the PDP context is successfully activated.

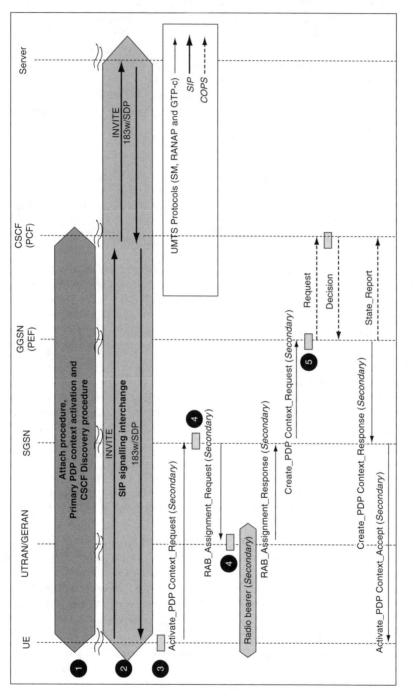

Figure 4.14 SIP session establishment flowchart [21]

In case of 'non-SBLP' session, even though there is no end-to-end QoS, there is still QoS up to the GGSN edge. In this case, GGSN performs DiffServ processing of the IP flows using the TFT provided by the UE at PDP context activation.

4.3.4 3GPP versus 3GPP2 in QoS

Once described the QoS mechanisms that are specified in 3GPP, we will have a look at the existing support for the provision of QoS given by 3GPP2, which is the standard body in charge of CDMA2000.

Currently, the CDMA2000 network architecture shows several shortcomings that prevent supporting multiple classes of traffic at the RAN, as it is done in 3GPP. However, some architectural modifications are today under study to be able to provide end-to-end QoS in the future and reach the goal of harmonizing 3GPP/3GPP2 IMS network architectures.

To illustrate how these shortcomings appear in the CDMA2000 network, we will describe first its architecture and the basics of the packet data service deployed over this network. Figure 4.15 shows the current CDMA2000 architecture without QoS support, its components and the protocol stack at each of them. The MS consists of two components: Terminal Equipment (TE) like for example a laptop, and Mobile Terminal (MT) such a PCMCIA CDMA2000 card. The CDMA2000 Base Station (BS), which consists of the Base Transceiver Station (BTS) and the Base Station Controller (BSC), is connected to the MS through the air interface. The BS together with the Packet Control Function (PCF) represents the RAN that connects the MS to the Packet Data Serving Node (PDSN), which works as a gateway to the IP network (Internet) [22].

To establish a data connection towards Internet, it is needed to set up first a PPP connection [23] between the TE and the PDSN. Data packets are transported from the TE to the PDSN in the form of PPP frames and, in the opposite direction, packets from the Internet received in the PDSN towards the TE are transported over the same PPP connection. A key issue of this

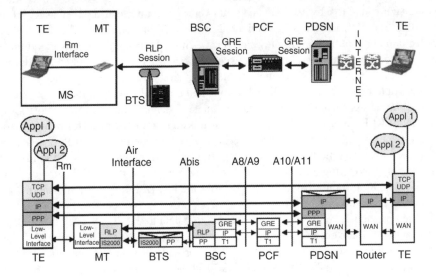

Figure 4.15   CDMA2000 architecture

way of operation is that the traffic for all applications running on the TE goes through the same PPP connection between the TE and the PDSN. This operation mode means a barrier for the QoS provision, as once the application packets have been encapsulated in PPP frames, it is impossible to distinguish those packets and handle them depending on the application QoS needs.

PPP frames received at the MT (in UL, or reverse direction) and at the BSC (in DL, or forward direction) are transported over a Radio Link Protocol (RLP) session to the BSC and MT respectively [24]. There is one RLP session per PPP session. Data sent through the RLP in either direction, is segmented into RLP frames before it is transmitted through the air interface.

The main functionality of the PCF is to direct PPP connection request from the TE to the appropriate PDSN that will handle connection. To carry the traffic between the BSC and the PCF a Generic Routing Encapsulation (GRE) [25] tunnel is used per PPP session. The GRE frames between the BSC and the PCF are normally transported over a private network, for example an IP network. Data between the PCF and the PDSN are transported over another set of GRE tunnels, again one per PPP session. The network that connects the PCF with the PDSN could be the public IP network.

To provide end-to-end QoS, some QoS parameters must be agreed at application level and mapped down onto the lower layers, i.e. link and network layers. Although the QoS parameters negotiation at application level could be performed with the existing network architecture by using SIP and SDP (an example can be found in previous section), the current link and network layers cannot support multiple RLP sessions and GRE tunnels to transport packets of different services in a single PPP session, i.e. there is no way of converting application level QoS parameters into link and network QoS parameters just because they do not exist. That is why some architectural modifications would be necessary in order to implement QoS mechanisms at the lower layers.

To make the shortcomings of the CDMA2000 network clearer, let us consider the case of a TE sending data to the PDSN. When a TE sends PPP frames to the other end, it could perform some kind of scheduling and prioritize in certain way those packets coming from applications with stringent QoS needs. This procedure could be carried out by classifying packets according to TCP or IP information, for example source TCP port. But once the PPP frames are converted to a byte stream and sent over the Rm interface and then as RLP frames to the BSC, these RLP frames are delivered at the BSC in sequence. This is ensured by sequence numbers on the RLP frames. Further, when the byte stream is converted into GRE frames and sent over the GRE tunnels between the BSC and the PCF and then between the PCF and the PDSN, these GRE frames are also delivered in sequence at the delivery points. So, there would be no way of handling real-time and non-real-time traffic that are part of the same RLP frame. As each RLP session has parameters such as number of retransmissions that applies to all RLP frames in the session, retransmissions are performed for real-time traffic although it does not make sense. Finally, because of ordered delivery of RLP, if a RLP frame that carries non-real-time traffic is lost, real-time RLP frames that follow the lost frame cannot be delivered to higher layers even if they are received at the BSC, i.e. the delivery of these real-time RLP frames can only happen after the lost RLP frame is retransmitted and is subsequently received at the BSC.

As stated previously, in the case of the TE sending data to the PDSN, RLP frames that are received at the BSC are encapsulated in GRE frames and sent to the PCF. Because GRE frames are also delivered in sequence, there is no way to provide service differentiation within the IP network that carries traffic between the BSC and the PCF. The same applies within the IP network that interconnects the PCF and the PDSN, as another GRE tunnel has to be used. To sum up, no QoS differentiation can be performed if all data to and from a TE is mapped onto a single RLP session and then onto a single GRE sessions.

The usage of multiple RLP sessions and multiple GRE tunnels turns essential to carry traffic to and from a TE supporting service differentiation. Currently, several standards have been released in which the requirements for the support of E2E QoS are specified. Those new standards introduced a new concept, the service instance. This new standard proposes that different PPP frames are sent in one single PPP session, each carrying a different class of service, over different service instances between the MT and the BSC [26]. Each service instance is implemented by a separate RLP session. Between the BSC and the PCF and between the PCF and the PDSN, there are multiple GRE sessions per TE, each of which carries a different class of traffic, that are also introduced in these standards.

Once each single class of traffic is carried in a separate RLP session, RLP sessions can be used to prioritize service classes between the MT and the BSC. Similarly, different GRE sessions are used to handle different classes of traffic and differentiate services.

The QoS support in the CDMA2000 wireless IP network could be then provided via one or more instances of a packet data service. The radio resources would be allocated per service instance and a service instance could request specific radio link QoS attributes. The types of these instances are defined as a main service instance or an auxiliary service instance, which reminds of the primary PDP context and secondary PDP contexts concepts in 3GPP. As a utilization example, we could say that the purpose of a main service instance is the provision of resources in the CDMA2000 wireless IP network to meet the QoS requirements for the applications that may require only BE QoS support. However, to meet the QoS demands of applications that require better than BE QoS, an auxiliary service instance can be used. The resource allocation for an auxiliary service instance is selective and is based on a characterization of QoS requirements associated with an application. One or more auxiliary service instances may be established by the MS based on the number of applications used for an MS each requiring different QoS.

On the one hand, the main service instance is a packet data service instance that is set up during the initial establishment of a packet data service. This packet data service instance normally has default QoS characteristics that are based on subscriber's profile and local policy. On the other hand, the auxiliary service instance is a packet data service instance that is set up on-demand to support a required QoS greater than the default QoS characteristics that are configured for the main service instance. This packet data service instance has QoS characteristics that are based on the request of the user, limited by the subscriber's QoS profile and local policy. With the introduction of the service instance, the 3GPP2 QoS support would become similar to the 3GPP Rel'99 one, according to the set of services that can be provided (background, interactive, streaming and conversational).

The latest standards released by 3GPP2 are already starting to specify some requirements and reference models for the provision of E2E QoS based on the service instance concept [27].

4.4 QoS Policy Management

The management of complex systems whose behaviour is sensibly affected by multiple factors and which are controlled through a plethora of parameters requires a management method that offers a simple way to control the overall system, but at the same time it enforces a consistent behaviour throughout the network.

A policy is essentially a matter of allocating resources in terms of business decisions. It is the translation between business terms and the configuration details necessary to produce those resource allocations what distinguishes policy management from configuration management. This process is further described along this section.

4.4.1 Motivation for QoS Policy Management

Internet-technology-based networks are being used for more functions and by more businesses. The users' ability to do business is affected by the health and abilities of their networks. As networks grow, the amount of different issues that need to be managed is also growing, not only because there are more devices to be managed, but also because the number of requirements needed to provide acceptable service quality increases (e.g. capabilities, number of services, types of interfaces, security, etc.).

As support for more functionalities is introduced in the network, the IT administrators must learn to use the management interfaces to manage the system. In addition, many of those management tools work with individual devices, so an administrator has to multiply the actions needed to configure one equipment by the number of similar devices in their network, even if they are of the same type and from the same vendor. The problem is exacerbated if the devices are from different vendors, since they must perform different tasks to manage similar capabilities. The same problem exists not just for networking configuration, but also for just about anything an IT administrator may need to manage.

In response to this situation, customers (IT administrators) have for many years been asking vendors for tools which better address their needs in managing such large and dynamic environments. Their list of desired features typically includes:

- Through centralized management, from a centralized control point, IT administrators may manage all network resources according to high-level rules.
- Abstracted or simplified management data, which is mapped to a detailed and proper management data across the network.
- Commonality across devices and consistency across interfaces.
- Automation of management tasks, which also minimize the OPEX.
- Fewer interfaces towards the huge diversity of network elements to be managed.

Policy management is seen as the answer to this problem. Basically, it offers a centralized control point, which administers the network in order to achieve consistent network behaviour for the QoS need of specific traffic flows.

This section will provide a background about the policy-based management, beginning with the evolution of the definition along standards, continuing with the description of the IETF Policy Model. Afterwards, the application of the QoS policy-based management to mobile networks is presented.

4.4.2 History and Evolution

The major effort for defining an overall framework for representing, deploying and managing policies is being developed by the IETF Policy Framework Working Group, as extensions to the Common Information Model (CIM) activity in the Distributed Management Task Force (DMTF). The CIM is mainly based on the definition of an object-oriented information model for representing all needed policy information.

The main premise in the definition of the Policy Core Informational Model is that it must have a generic nature and be applicable either to QoS, to non-QoS networking applications (e.g. DHCP and IPSec), or to non-networking applications (e.g. backup policies, auditing access, etc.).

As aforementioned, policies represent business goals and objectives. Thus, a translation must be made between these goals and objectives and their realization in the network. Not only the high-level descriptions of network services and metrics must be translated into lower-level, but also vendor- and device-independent specifications. The Policy Core Information Model classes are intended to serve as the foundation for these lower-level, vendor- and device-independent specifications.

Some examples of areas in which the Policy Management has been extensively used are the resource administration and security. As to QoS Policy Management, the most advanced model has been developed for DiffServ management [8] and MPLS Traffic Engineering [11].

4.4.2.1 DiffServ Management Model

In the DiffServ[4] case, a management model of DiffServ routers for use in their management and configuration is defined. DiffServ functionalities for example classifiers, meters, actions (e.g. marking, absolute dropping, counting, multiplexing), algorithmic droppers, queues and schedulers may be managed by high-level policies from a centralized control point. It describes possible configuration parameters for these elements and how they might be interconnected to realize the range of traffic conditioning and PHB functionalities described in the DiffServ architecture [28]. The model is intended to be abstract, and representing the main configuration parameters to DiffServ functionality.

4.4.2.2 MPLS Management Model

Among the applications of policy management to MPLS Traffic Engineering, perhaps the most interesting one relates to MPLS/DiffServ traffic monitoring of point-to-point MPLS/DiffServ tunnels, as the monitoring parameters are performance indicators that will facilitate operators to conduct traffic engineering, service assurance and performance management. The MPLS/DiffServ traffic monitoring parameters can be divided into two groups: MPLS/DiffServ LSP performance parameters and MPLS/DiffServ network performance parameters. The former concerns the performance of individual LSP and the latter deals with the performance of an MPLS network.

Every LSP is established to achieve pre-defined goals. The goals may be specified in an SLA. For instance, if an LSP is dedicated to carry customer application traffic, the LSP must meet the SLA. The pre-defined goals may also be set by an operator. For instance, if an LSP is used to route control and management traffic within a network, it must meet the performance objectives set by the operator. To ensure that MPLS/DiffServ LSPs in a network achieve

---

[4] DiffServ was described in section 4.2.

their goals and that proactive actions are taken when the goals are not met, the MPLS/DiffServ traffic-monitoring framework is managed through a policy-based model.

4.4.3 IETF Policy Model

4.4.3.1 Policy Architecture and Definitions

A policy infrastructure is the set of protocols, information models and services that allow the network administrator to translate high-level requirements based on expected quality and business premises into different treatment of packet flows.

Policy management is based on rules containing conditions and actions. For example, an operator could configure the network in order to provide the highest priority to those packets which belong to a '*gold*' user profile, or even to packets which belong to a real-time voice service, independently of the user. This rule-based management system resembles the operational principles of the IETF policy control framework. This policy model is composed of three types of entities (as depicted in Figure 4.16).

1. *Policy Repository* – is the location containing high-level policies defined by the administrator that can be applied within the policy domain.
2. *Policy Decision Point (PDP)* – represents a group of functions in charge of acquiring, deploying, and optionally translating policy rules into a form understandable by a PEP.
3. *Policy Enforcement Point (PEP)* – represents the entity whose behaviour is going to be managed by the policy rules.

Exchange of information between different types of entities needs the use of specific protocols capable of transmitting the policy information. The Common Open Policy Service (COPS) protocol is a client–server protocol intended for the communication of policy requests and decisions between a PEP and a PDP. COPS is characterized by its reliability and fault-tolerance, unlike legacy control protocol such as Simple Network Management Protocol

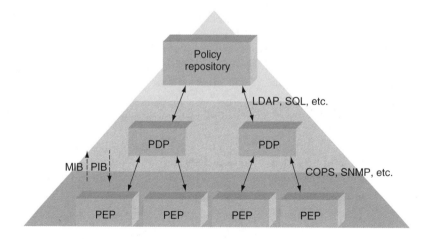

Figure 4.16   IETF policy architecture

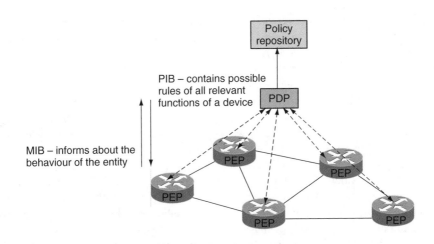

Figure 4.17   PIB and MIB definition

(SNMP). Furthermore, COPS is flexible and easy to use for both dynamic and static configuration. Depending on the type of policy repository, the access can be performed by different
protocols such as Lightweight Directory Access Protocol (LDAP) or SQL.

Finally, network devices must be configured with the policy information defined in
the policy rules. Policy Information Base (PIB) defines the data structure by which the PDP
downloads policy into a PEP. Each PIB contains the possible rules of all relevant functions of
a specific device.

In addition, Management Information Base (MIB) has already been defined with SNMP
protocol in IETF, for the purpose of monitoring and allow simple configuration of network
devices, before the introduction of a policy framework. MIB defines the device-specific data
structure by which the PEP may inform its behaviour to the PDP (Figure 4.17).

4.4.3.2 Policy Modelling

The main goal of the policy information model is to help building a bridge over the conceptual
gap between a human policy maker and the capabilities of the network element that is configured
to enforce the policy. Clearly, this wide gap implies several translation levels, from the abstract
level to the implementation level. At the abstract level are the business QoS policy rules. The
Information Model facilitates a formal representation of network QoS business rules, thus
providing the first concretization level: formally representing humanly expressed QoS policy.

When a human business executive defines the network policy, it is usually done using
informal business terms and language. For example, a human may utter a policy statement
that reads: 'traffic generated by our human-resources application should have higher priority
than traffic generated by people browsing the WEB during their lunch breaks'. While this
statement clearly defines QoS policy, the network itself cannot enforce it. A translation to
'network terms and language' is required.

The core of the policy management is the system modelling from the management point of
view. For that purpose, an object-oriented information model is developed for representing
the policy information in the PIB.

One way to think of a policy-controlled network is to first model the network as a state machine and then use the policy to control in which state a policy-controlled device should be, or is allowed to be, at any given time. Given this approach, a policy is applied using a set of policy rules. Each policy rule consists of a set of conditions and a set of actions. Policy rules may be aggregated into policy groups, and these groups may be nested, to represent a hierarchy of policies.

If the set of conditions associated with a policy rule evaluates to 'true', then a set of actions that either maintain the current state of the object or force a transition to a new state may be executed. For the set of actions associated with a policy rule, it is possible to specify an order of execution, as well as an indication of whether the order is required or merely recommended. It is also possible to indicate that the order in which the actions are executed does not matter.

The scope of the policy rules can vary. *Global policies* have a global effect on the end result; i.e. the end result of the global policy relies on being executed globally across the entire path of the network service. *Domain policies* are meaningful within one domain of QoS mechanisms, for example radio domain. *Local policies* are only meaningful within one network element. From the viewpoint of rule management, it is essential that the global and domain policies are harmonized. This target is easily achieved by means of a central control point. Local policies can reside within the control of a local element, as long as such local policies have no effect on the QoS globally – mainly those whose decision-time response is critical.

Policy groups and rules can be classified by their purpose and intent. It indicates whether the policy is used to motivate when or how an action occurs, or to characterize services (that can then be used, for example, to bind clients to network services). The following describes each of these concepts in more detail.

- Motivational Policies are solely targeted at whether or how a policy's goal is accomplished. Configuration and Usage Policies are specific kinds of Motivational Policies. Another example is the scheduling of file backup based on disk write activity from 8 am to 3 pm.
- Configuration Policies define the default (or generic) set-up of a managed entity (e.g. a network service). Examples of Configuration Policies are the set-up of a network forwarding service or a network-hosted print queue.
- Installation Policies define what can and cannot be put on a system or component, as well as the configuration of the mechanisms that perform the install. Installation policies typically represent specific administrative permissions, and can also represent dependencies between different components (e.g. to complete the installation of component A, components B and C must be previously successfully installed or uninstalled).
- Error and Event Policies. For example, if a device fails between 8 am and 9 pm, call the system administrator, otherwise call the Help Desk.
- Usage Policies control the selection and configuration of entities based on specific 'usage' data. Configuration Policies can be modified or simply re-applied by Usage Policies. Examples of Usage Policies include upgrading network forwarding services after a user is verified to be a member of a 'gold' service group, or reconfiguring a printer to be able to handle the next job in its queue.
- Security Policies deal with verifying that the client is actually who the client purports to be, permitting or denying access to resources, selecting and applying appropriate authentication mechanisms, and performing accounting and auditing of resources.
- Service Policies characterize network and other services (not use them). For example, all wide-area backbone interfaces shall use a specific type of queuing.

4.4.4 Policy Management in Mobile Networks

Telecommunications mobile networks are a clear example of a complex system which is suffering a big expansion over the world in the sense of major diversity of devices and services. Starting from a BE service model, future mobile networks are intended to support different service levels for specific QoS requirements.

This growth in network usage and technology complexity has caused a number of management-related problems on many areas, including the control of the QoS. The main problem for operators is how to manage these networks in a simple way, and at the same time offer the possibility to regulate access to network resources based on different criteria, such as user profiles or applications.

After presenting the IETF policy architecture in the previous section, the QoS framework in UMTS is outlined to present the integration of the policy architecture into the UMTS framework. The QoS functions, which should be supported in UMTS in order to apply QoS differentiation, are numerous especially if we assume that the Internet Protocol (IP) will be widely used at the transport layer. At the UMTS QoS level, the required functions can be listed briefly: subscription control, authorization, admission control, resource reservation, resource monitoring, link adaptation, bearer-level classification, shaping and radio scheduling. At the IP transport level, the DiffServ management model specified in the IETF can be reused. This model includes classification, metering, marking, scheduling and a dropping mechanism. If other transport technologies such as MPLS are going to be applied, the complexity of the QoS system will increase even more.

Policy can be used for defining the behaviour of different elements and functions of the network.

- *The Gi reference point*: This includes APN configuration, DiffServ marking, and establishment of connection to ISP corporate (tunnelling, etc.).
- *The GGSN PHB*: This includes configuration of the queue weights and maximum amount of real-time traffic allowed.
- *The GPRS backbone*: This includes configuring DiffServ marking at the edge (RNC, GGSN, etc.), and managing PHB of routers.
- *PHB of other nodes*: Although SGSN and RAN elements are normally less linked to services than GGSN, it would be possible to manage their QoS-related functionalities based on high-level rules.
- APN-improving GGSN QoS configuration.
- Setting QoS based on services.

In a nutshell, the Policy Management provides a tool for the network administrator to offer a variety of QoS network services to different customer needs and for different traffic flow types. Furthermore, it allows a dynamic management of the network, providing a consistent control of rules set across the network for the desired QoS levels in the network.

References

[1] R. Braden *et al.*, 'Integrated Services in the Internet Architecture: An Overview', RFC 1633.
[2] S. Shenker *et al.*, 'Specification of Guaranteed Quality of Service', RFC 2212.

[3] J. Wroclawski, 'Specification of the Controlled-Load Network Element Service', RFC 2211.

[4] R. Braden *et al.*, 'Resource reSerVation Protocol – Version 1 Functional Specification', RFC 2205.

[5] J. Wroclawski, 'The Use of RSVP with IETF Integrated Services', RFC 2210.

[6] K. Guo, S. Rangarajan, A. Siddiqui and S. Paul, 'Providing End-to-End QoS for Multimedia Applications in 3G Wireless Networks', Center for Networking Research, Bell Laboratories, Lucent Technologies, Holmdel, NJ, USA, http://www.bell-labs.com/user/kguo/papers/qos-spie03.pdf.

[7] K. Nichols *et al.*, 'Definition of the Differentiated Services Field (DS Field) in the IPv4 and IPv6 Headers', RFC 2474.

[8] S. Blake *et al.*, 'An Architecture for Differentiated Services', RFC 2475. ⁻

[9] V. Jacobson *et al.*, 'An Expedited Forwarding PHB', RFC 2598.

[10] J. Heinanen *et al.*, 'Assured Forwarding PHB', RFC 2597.

[11] D. Awduche *et al.*, 'Requirements for Traffic Engineering over MPLS', RFC 2702.

[12] 3GPP TS 23.207, 'End-to-End Quality of Service (QoS) Concept and Architecture' (Release 6) V6.3.0, June 2004.

[13] S. Floyd and V. Jacobson, 'Link-sharing and Resource Management Models for Packet Networks', IEEE/ACM Transactions on Networking, Vol. 3, No. 4, pp. 365–386, August 1995.

[14] D. Fernãndez and H. Montes, 'En Enhanced Quality of Service Method for Guaranteed Bitrate Services over Shared Channels in EGPRS Systems', IEEE Vehicular Technology Conference, Spring 2002, Birmingham, USA.

[15] Cisco Web page, http://www.cisco.com/warp/public/732/netflow/qos_ds.html.

[16] J. Halonen and Romero J. Melero, 'GSM, GPRS and EDGE Performance, Evolution Towards 3G/UMTS', 2nd Edition, Ed. Wiley, 2003.

[17] 3GPP TS 23.107, 'QoS Concept and Architecture' V6.1.0.

[18] 3GPP 23.060, Release 1999: 'General Packet Radio Service (GPRS), Service Description, Stage 2'.

[19] M. Handley, H. Schulzrinne, E. Schooler and J. Rosenberg, 'SIP: Session Initiation Protocol', IETF RFC 2543, March 1999.

[20] M. Handley and V. Jacobson, 'SDP: Session Description Protocol,' IETF RFC 2327, 1998.

[21] H. Montes, G. Gómez, R. Cuny and J. F. Paris, Deployment of IP Multimedia Streaming Services in Third Generation Mobile Networks, *IEEE Wireless Communications magazine*, Special Issue on IP Multimedia Services in Next Generation Mobile Networks, October 2002.

[22] K. Guo, S. Rangarajan, A. Siddiqui and S. Paul, 'Providing End-to-End QoS for Multimedia Applications in 3G Wireless Networks', Center for Networking Research, Bell Laboratories, Lucent Technologies, Holmdel, NJ, USA, http://www.bell-labs.com/user/kguo/papers/qos-spie03.pdf.

[23] W. Simpson, 'The Point-to-Point Protocol (PPP)', RFC-1661, July 1994.

[24] 3GPP2 C.S0017-0-2, 'Data Service Options for Spread Spectrum Systems – Addendum 2', V2.0, August 2000.

[25] S. Hanks, T. Li, D. Farinacci and P.Tranina, 'Generic Routing Encapsulation', RFC 1701, October 1994.

[26] 3GPP2 P.S0001, 'Wireless IP Network Standard', October 2002.

[27] 3GPP2 S.R0079-0, 'Support for End-to-End QoS', May 2004.

[28] M. Carlson, W. Weiss, S. Blake, Z. Wang, D. Black and E. Davies, 'An Architecture for Differentiated Services', RFC 2475, December 1998.

5

End-to-End Service Performance Analysis

Rafael Sánchez, Gerardo Gómez, Pablo Ameigeiras, Jorge Navarro and Gabriel Ramos

5.1 Introduction

Typically voice wireless service providers have based their measurements on general indicators like cell coverage and capacity in order to assess their network performance. Additionally, for the characterization of specific markets, measurements such as *drop call rate, block call rate* or *voice quality* were used. The main advantages of these measurements were that they could be easily taken through network monitoring, they are general enough to be used as comparison between different systems, and there is good correlation with the performance experienced by the voice user. However, these voice performance indicators are based on specific characteristics of Circuit Switched (CS) traffic, such as dedicated channels for each user and no error correction in case of frame loss.

When moving to the area of data services, traffic sources present very different and variable profiles. Users can share the resources of same data channel, the rate of the assigned channel may be changed according to user conditions, and retransmissions can be performed in case of transmission errors.

This variability in the traffic distribution makes it difficult to define common measurements which can give, at the same time, the performance of the network and the user perception of each service. A clear example can be seen in traditional network monitoring, which averages the performance of different traffic profiles (Web download, streaming, e-mail, etc.), by aggregating reported events over certain period. The typical reporting periods of hours, days or weeks, average not only over the time, but also over different services, file sizes, different burstyness, etc. The result is that the indicators obtained are so mixed that they lose the correlation with individual user experience.

End-to-End Quality of Service over Cellular Networks: Data Services Performance and Optimization in 2G/3G
Edited by G. Gómez and R. Sánchez   © 2005 John Wiley & Sons, Ltd

The complexity in the data measurements makes it necessary to produce a more detailed analysis which will have to differentiate between indicators of the network and user performance separately. However, the analysis should not be fully independent, but try to establish correlation points between the different measurements, which will help to understand better the evolution of both the network and the users.

Services and their associated metrics were already introduced in Chapter 3. In this chapter, the focus will be set on how the user performance can be characterized through the aggregation of the effect of the different protocol layers, taking into account not only the characteristics of the services but also the interactions that they may have with the architecture and protocols of the wireless network. In Chapter 6, the analysis will continue with the definition and classification of general performance indicators for data services, and provide practical examples of network analysis.

5.1.1 End-User Performance Analysis

The term *End-User Performance* is used to refer to the Quality of Service (QoS) that the user perceives when using an application. No matter which wireless network lies beneath the transmission (whether it is GPRS, EDGE, CDMA2000, WCDMA or WiFi), the metrics which define the performance of one application are directly linked to the characteristics of the service itself.

From the user point of view, the whole network can be considered as a data bearer which provides certain transmission capabilities. These capabilities can be defined by the combination of two basic parameters: *throughput* and *latency*. However, in order to fully define the access network it is also necessary to consider other additional factors, like the *accessibility* and the *retainability* of the connections. These terms can be described as follows.

- *Throughput* – is the amount of data transferred in one direction over a link divided by the time taken to transfer it, expressed in bits or bytes per second. Generally the throughput in the radio interface is measured as the effective Radio Link Control (RLC) payload throughput. This indicator depends on many factors, such as the terminal capability, cellular planning, load conditions or radio technology.
- *Latency* – is the time it takes for a packet to cross a network connection, from sender to receiver. Latency is particularly important for a synchronous protocol where each packet must be acknowledged before the next can be transmitted. For example, TCP establishments and congestion control algorithms are very much affected by the latency of the network, as described in Chapter 3.
- *Accessibility* – is the probability that the service is available when user requires it. In the case of a wireless network it will depend on several factors, from the signal coverage to the network connectivity with external nodes and application servers. The access to a service may fail for any cause, some of them inside the operator's network, but some others by the external network itself (Internet). A wireless network operator shall take care of the optimization of the part of the transmission that goes through its network: radio accessibility and session accessibility.
- *Retainability* – of the network will be given by the probability that data connections are able to finish successfully, without transmission errors that may affect the performance. From the operator point of view there are different levels of retainability: radio retainability and session retainability.

However, from the user point of view, still problems with radio accessibility and retainability are finally seen as throughput or delay degradation if the radio network is able to recover transparently to the user from establishment failures and drops.

The data bearer provided by the network is directly connected to the Internet transmission protocols, and is used to create an IP pipe that will be used to transmit the transport and application layer data. Figure 5.1 represents this IP pipe as a tube where the diameter is determined by the bandwidth and the length as the delay of the data bearer.

This separation between application and transport protocols, which are service dependent, and the data bearer, which depends on the transmission technology being used, is the main breakdown for the analysis of a certain access network. From IP level and above, the protocol effects and degradations that may affect to the user are the same regardless of the underlying system, and in most of the cases this performance can be characterized just based on the size of the IP pipe that is provided and the round-trip delay between client and server.

This is true in most of the cases, although there are certain cellular network effects that can also affect the variability of the IP pipe. This means that depending on the radio conditions or other network-related factors, the IP pipe may keep with a constant behaviour during a whole connection, or on the contrary, it may present certain variations. These variations can be of two types: slow variations, which may be due to variability on the radio conditions affecting the average throughput; or sudden variations, which will be seen as step changes in the performance of the pipe. These step changes can be due to certain cellular events (such as cell reselections, radio bearer upgrades, shadow reception areas, etc.) that have specific impacts on the performance of the protocols.

One typical situation is a TCP transmission which adapts the IP transmission rate to the Bandwidth Delay Product (BDP) of the system (see section 5.4.1.2 for further details on BDP concept). Under this scenario, some radio specific factors (e.g. cell reselection, shadow reception area) may interrupt abruptly the IP pipe for few seconds, and re-establish after that. The Radio Network is able to recover from such a situation by buffering the user data till the radio connection is re-established, but very long gaps may cause TCP retransmission timer to expire and start sending again the whole transmission window, even if the data is not lost, but only delayed. This effect is known as spurious retransmissions [1] and will be described in depth in section 5.4.1.4.

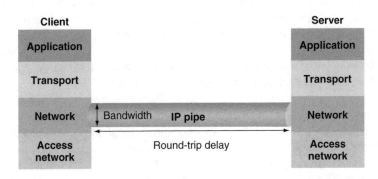

Figure 5.1   IP data transmission pipe view

The variability of the wireless network bearer does not follow the same rules in different systems, so a deep understanding of the system is necessary in order to understand what effects are more sensitive from the user point of view. Optimization of the radio network will have as an objective to provide bearers with higher throughputs, lower delays and better stability. In addition, this information can be also used from application and transport point of view to configure the behaviour of the protocols in such ways that can react better to these variations.

One common problem for operators is how to set reasonable expected performance levels for the different services that are supported in their networks. The theoretical radio capacities or even the measured network indicators can be always used as a reference; but when it comes to a measure of what would the user be expected to see, many other factors shall be taken into account. A thoroughly understanding of the mechanisms of the different protocol layers involved in a data session and the delays and packet losses that may affect during the transmission is very important for the people in charge of monitoring this performance.

Once all the factors, or at least the most significant factors, which contribute to the service-level degradation of one user are understood, it would be possible to build simple models which are able to provide estimations of the target thresholds that could be expected from a user point of view. If the models are simple enough and based on certain common metrics for different transmission technologies, they can also provide a simple way to forecast the user-level performance after an improvement in any of these metrics, or to provide benchmarking references for multiple technologies.

5.2 Service Performance Characterization

Legacy cellular networks do not implement a full end-to-end QoS approach. They operate just as an additional node integrated in the whole Internet, where the access point to the wireless users is a router (the GGSN) which separates the Internet domain from the operators' domain. Inside the operators' domain, a series of mechanisms aimed to provide QoS assurance may be implemented according to the 3GPP and 3GPP2 specifications (as described in Chapter 4), but part of the performance will depend on the implementation of the external network and the protocols that are used. The interface to the external network from the layering point of view is the IP level, which will be affected by routing policies and capacities, but the control of the transport and application layers will depend very much on the configuration of the end peers, typically client and server, and the version of the protocol used.

According to this division in operator domain and external domain, the analysis of the service performance can be done from two perspectives (Figure 5.2).

1. The *Quality of Service (QoS)* is focused on how the service is performing from a network point of view. The QoS is defined by mechanisms and procedures providing traffic differentiation and assuring certain quality requirements.
2. The *Quality of Experience (QoE)* focuses on how a service is performing from the user point of view. This kind of approach sees the service as a whole and takes care of every possible aspect which would affect the performance.

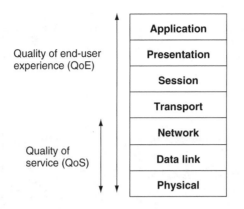

Figure 5.2   Service performance analysis scope

In Chapter 4, different approaches to provide QoS have been already described. In this section, the focus will be set on the end-user experience, trying to understand the different factors contributing to its degradation, and defining simple models that can show the performance over different services.

5.2.1 Characterization of End-User Performance

The end-user performance is affected by every protocol layer and network element in the connection path, from one user terminal to other user or server in the remote end of the network. Based on this consideration, a way to analyse and estimate the end-user experience has been defined. This method follows a bottom-up approach, starting from the lower levels of the layer architecture and considering a cumulative degradation of the performance based on the effects of the different layers and their interactions. The ideal throughput provided by layer one (physical layer) is considered initially as the starting point, and then the performance degradation introduced by each of the upper layers in the protocol stack is estimated.

The way in which the performance of a certain service is degraded, compared to others, will depend also on the kind of service and the factors towards it is more sensitive: throughput, delay, response time, packet loss, etc.

Figure 5.3 presents, as an example, the different protocol stack layers in GERAN networks. But similar concept can be applied to other technologies also.

From a service performance point of view, the set of factors that produce a degradation of the link level throughput can be grouped into data link level and upper layer effects.

- *Data link effects* are those factors affecting the performance that depend on the network conditions, such as interference, coverage, radio resource sharing, etc. The resulting performance after these effects is the *data link throughput*, which refers to the payload (throughput) offered by the Radio Access Network (RAN) to the transport and application

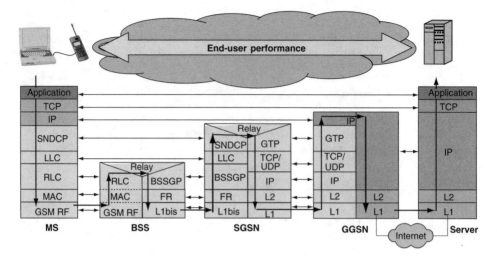

Figure 5.3 Layers affecting end-to-end performance (GERAN)

layers. Data link throughput and the network latency are the basic elements that can be calculated for different networks and used as a starting point to estimate the performance of each service. Both measurements depend only on the network itself. Data link effects are further described in section 5.3.

- *The upper layer effects* are those factors which depend on the transport and application protocols. They contribute to the degradation of the user service performance in the same way, independently on the network that is used for the transmission. What determines the proportional impact of these effects on the final service performance are both the throughput and the latency at the data link level. Normally networks with higher latencies will suffer much more from these degradations than networks with less latency, even if the data link layer throughputs are quite similar. Upper layer effects are deeply analysed in section 5.4.

In case of services based on *TCP* transport protocols like Web browsing or *File Transfer Protocol* (*FTP*), the main effects are due to header overheads, *TCP* connection establishment and slow start, specific cellular events over *TCP* layer and, finally, application layer protocol. Other transport layers, such as *User Datagram Protocol* (*UDP*), which introduce less control mechanisms and do not require so much interaction between transmitter and receiver will have a more reduced impact on the overall end-user performance but at the cost of not giving transmission reliability of network congestion mechanisms.

Figure 5.4 illustrates the evaluation method for the particular case of services based on TCP.

In Figure 5.5, a numerical example of FTP performance degradation introduced by the aforementioned factors in GPRS (3 + 1 TSL terminal capability) is depicted. Note that in this example (10-kB file size) the application throughput is reduced to a half compared to the ideal RLC throughput in the radio interface.

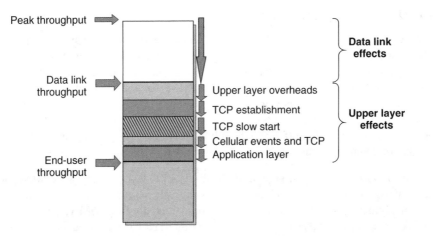

Figure 5.4   Main sources of throughput degradation

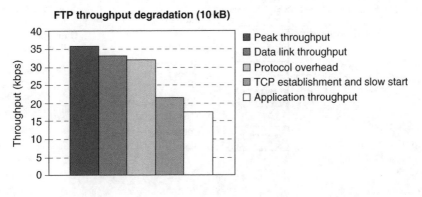

Figure 5.5   Example of performance degradation for GPRS

The following sections are aimed at providing a description of the different factors affecting the end-to-end performance.

5.3 Data Link Effects

As aforementioned in the previous section, the maximum theoretical throughput that could be delivered to a mobile terminal is degraded by a set of factors that mainly depends on the network configuration, load conditions, radio link quality and the adopted Radio Resource Management (RRM) strategy.

A proper analysis of these effects must be treated technology by technology. In this book we analyse the (E)GPRS and WCDMA cases as examples of the two main families of radio access networks: Frequency and Time Division Multiple Access (FDD + TDMA) and Code Division Multiple Access (CDMA).

5.3.1 Data Link Effects in (E)GPRS

(E)GPRS, as described in section 2.2, is a GSM-based technology which is affected not only by the interference levels derived from the frequency planning (i.e. number of frequencies, frequency hopping, etc.), but also by the delays introduced by the network architecture itself (i.e. transmission times between BTS and BSC), RRM functions and radio protocol functionalities. The main sources of performance degradation in this particular case of (E)GPRS networks are depicted in Figure 5.6.

Focusing on the data link effects and applying a bottom-up approach from physical to IP layer, the following performance indicators can be defined.

Peak throughput: It is the throughput delivered to LLC layer without RLC/MAC headers, which depends on the used Modulation and Coding Scheme (MCS). For instance, for GPRS CS-2, maximum gross throughput is 13.4 kbps (including RLC/MAC headers), where 12 kbps corresponds to the peak air throughput without RLC/MAC headers. The radio block structure for GPRS and EGPRS is illustrated in Figure 5.7.

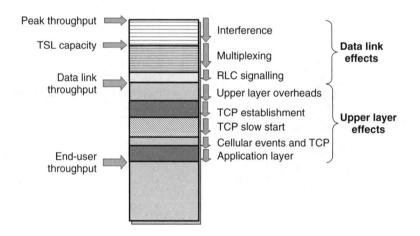

Figure 5.6   Main sources of throughput degradation in (E)GPRS

Radio block			
MAC header	RLC header	RLC data	BCS

Radio block structure for data transfer for GPRS

Radio block			
RLC/MAC header	HCS	RLC data	BCS

Radio block structure for data transfer for EGPRS

Figure 5.7   Radio block structure for GPRS and EGPRS

Timeslot (TSL) capacity: It is the available throughput in a fully utilized TSL after including the effects of interference, also including RLC retransmissions if RLC acknowledged mode is used. TSL capacity depends on different factors, like: network planning (frequency reuse) and configuration, radio link quality, layer where GPRS is allocated (BCCH, hopping, non-hopping) and Effective Frequency Load (EFL).

Reduction Factor (RF): This concept includes the impact of the resource sharing among different connections, i.e. the available air throughput per user after sharing the radio TSLs between several connections. The RF is conditioned by the network load and the dimensioning criteria, and depends on the following factors.

- *GPRS allocation size*: The number of reserved TSLs for (E)GPRS and shared TSLs between voice and data is very important to prevent high RFs and GPRS blocking.
- *CS load and pre-emption criteria*: Depending on the pre-emption criteria used, CS traffic may or may not pre-empt Packet Switched (PS) traffic from shared TSLs. If CS traffic has higher priority than PS traffic, high CS load will also lead to a high RF for PS traffic.
- *Terminal capability (number of TSLs supported)*: Terminals with higher TSL capability are able to get higher bit rates from the system, and also have a higher probability of radio resource sharing among other terminals. However, statistical multiplexing gains [1] make it more beneficial to have more TSLs, even if that means higher multiplexing levels.
- *RRM scheme*: The RRM takes care of minimizing the TSL sharing when doing channel allocations, ensures that a terminal is connected to the best cell, and may also support QoS mechanisms (like QoS-aware scheduler) that will prioritize certain flows at the expense of queuing others with less importance. Hence, flows with higher priority will utilize the radio resources for a longer period of time (in average).

The concept of TSL sharing is illustrated in the Figure 5.8. Practical examples with RFs values are also provided in section 5.5.

RLC signalling: (E)GPRS requires the establishment of a Temporary Block Flow (TBF) at RLC level whenever any data need to be sent through the radio interface.[1] Taking into account that the TBF establishment procedure takes certain time (from 300 to 600 ms), continuous TBF establishments and releases may produce performance degradation. The RLC control

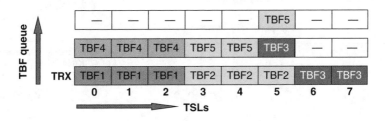

Figure 5.8   TSL sharing in (E)GPRS

---

[1] A TBF is established whenever new incoming data needs to be sent over the Um (radio) interface. Such TBF is released when RLC buffers are empty after all available data is sent.

blocks used for signalling in downlink direction also share the same radio resources as that of the data RLC blocks, thus they also have an impact to reduce the user downlink throughput (same applies to uplink). In GPRS, different procedures might require RLC control blocks transmitted in downlink. The most common case is the transmission of data in uplink, which requires at least one control block in order to assign the resources during the establishment of one uplink TBF. In case the uplink TBF is established when there is already a downlink TBF active, this assignment is sent in the PACCH, which is multiplexed with downlink PDTCH [2]. In RLC-acknowledged mode, it is also required to periodically send a number of control blocks, transmitted in the PACCH, carrying acknowledgements (ACKs) for uplink data. From the above description, it can be derived that this effect is highly dependent not only on the traffic characteristics, but also on the coding scheme that is in use. With higher coding schemes each control block means more potential lost bits for data transmission. The following formula tries to describe this degradation in a general case:

$$\left(1 - \frac{\text{Downloaded Bits Per Uplink TBF}}{\text{Bits Per Radio Block}(1 + \text{RLC Acks Per Uplink TBF}) + \text{Downloaded Bits Per Uplink TBF}}\right)$$

As commented before, the amount of signalling traffic will very much depend on the traffic itself. For instance, in the case of TCP download, the UL TCP ACKs of 40 bytes are regularly generated for every two downlink TCP data segments, introducing certain throughput degradation (Figure 5.9). On the contrary, for UDP-based traffic such as streaming, there are no transport layer ACKs, so the proportion of control block overheads will be lower in average. Depending on the traffic characteristics of a particular service, the need for establishing and releasing TBFs may also vary. For instance, for TCP traffic, TCP ACKs generate a very bursty traffic composed of small packets, leading to frequent uplink TBF establishments and releases.

In order to decrease the latency, some enhancements were introduced in (E)GPRS specifications: delayed downlink TBF release feature is available since Rel'97 and Extended uplink TBF mode feature is available in Rel'4 mobile terminals. Both features are based on the idea of not releasing the resources once there is not any more data to transmit. Instead, a timer is started and the TBFs are kept until this timer expires. When traffic is bursty it is quite usual that buffers get empty for small periods, and the time needed for establishing and releasing the TBFs could cause significant service performance degradation. The length of this timer shall be set to a suitable value in the range of 1 to 5 seconds according to specs [2]. Additionally, both features allow decreasing the average RTT of the network [3].

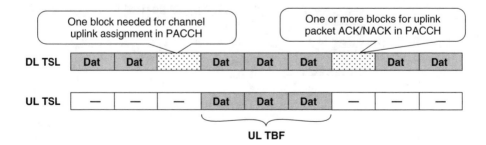

Figure 5.9   Effect of RLC control blocks on transmission

Apart from the effects listed before, there exist many other wireless events that directly or indirectly may affect upper layers' behaviour.

One of the most important ones in (E)GPRS comes from the mobility issue. Typical outage times[2] during cell reselections under (E)GPRS networks take from one to several seconds. This fact makes it impossible to support conversational services with certain quality, although real-time streaming services could still be supported, thanks to the use of buffers at application level which compensate such delay variations. In this respect, different mechanisms have been standardized to minimize this outage time, such as Network Controlled Cell Reselection (NCCR), Network Assisted Cell Change (NACC) or usage of Packet Common Control Channels (PCCCH) for signalling, which together may minimize those delays to around 500 ms.

Another important aspect is related to the retransmissions at RLC layer. Depending on the reliability requirements of the application, RLC-acknowledged or RLC-unacknowledged mode may be used. RLC-acknowledged mode provides high reliability in the transmission, but at the expense of higher delay variations due to RLC retransmissions. Generally, this is not a big problem for non-real-time services and even for some real-time services if they use a small buffer at the application. The benefits of reliable RLC transmission generally overcome the performance of unacknowledged mode.

In general, delay variations, wherever they come from, will impact upper layers' performance, especially TCP and application, which will be analysed in section 5.4.

5.3.2 Data Link Effects in WCDMA

The physical layer of WCDMA offers the service of transferring user data through various transport channels. These transport channels can be classified into three groups: common channels, Dedicated Channels (DCH), or Downlink Shared Channels (DSCH), as described in section 2.3.

Common channels are characterized by using explicit addressing of the mobile involved in the data transfer. The common channels that carry user data are: FACH, RACH and CPCH.

- The *Forward Access Channel (FACH)* is a downlink channel that is intended to carry control information to a mobile known to a location in the cell. The signal of the FACH must reach the entire cell, for all the mobiles to be able to decode its information. Additionally, the FACH can carry packet data information for any user in the cell.
- The *Random Access Channel (RACH)* is an uplink channel intended for signalling. The RACH is always received from the entire cell, and is characterized by a collision risk and by being transmitted using open-loop power control. It can also transmit user data.
- The *Common Packet Channel (CPCH)* is an uplink channel intended for packet data transmission. The CPCH could be considered as an extension of the RACH where its transmission may last for several frames. CPCH is associated with a DCH on the downlink, which provides fast power control and CPCH control commands.

---

[2] Outage time during cell reselection can be defined as the time from sending the last data block in serving cell until the reception of first data block in target cell (in case of ongoing data transfer).

As stated in section 2.3, there exists only one type of dedicated channel in WCDMA: the DCH. In the DCH, there is no need for explicit addressing of the target mobile. The DCH can carry all the information (i.e. user data and higher layer signalling) intended for a given user in both uplink and downlink. The DCH makes use of both fast power control and soft handover.

The DSCH is a downlink channel shared by several mobiles. All the resources assigned to this channel (i.e. power and code resources) are time multiplexed between all the users sharing the channel. The DSCH is associated with one or several downlink DCHs that are used to indicate which mobile will be targeted for transmission in every time interval. This channel makes use of fast power control although it does not use soft handover. The main advantage of the DSCH is that the reserved spreading code can be shared among all the users in the channel. The sharing of the code resources reduces their consumption, which improves the network performance in situations of code shortage. This is particularly interesting for non-real-time services due to the bursty nature of their traffic.

The bit-rate capabilities of the above-described channels significantly vary due to their different characteristics. The power efficiency of common channels is pretty low because they do not use fast power control (except the CPCH) and soft handover. Thereby, the bit rate associated with these channels is typically low (e.g. 8–32 kbps), and for this reason common channels are not used for transmission of large amounts of data. On the other hand, DCHs are much more efficient from a power transmission viewpoint, and they typically support a much wider range of bit rates (e.g. 8–512 kbps). However, the process of establishing a DCH brings an associated delay, hence, the transmission of small packets could be carried out faster over a common channel. The DSCH typically requires an allocation of large amount of power and code resources that enables high channel bit rates (e.g. 64–512 kbps). Note that high bit-rate capabilities must be divided between all the users sharing the channel. In any case, the power transmission of the DSCH is not as efficient as the DCH [5], and thereby its usage is only convenient for code shortage situations. The characteristics of the different channels for user data transmission are summarized in Table 5.1.

When a mobile is initially switched on, it has to select a PLMN (Public Land Mobile Network) to connect to. After the initial synchronization procedure, the mobile is in *idle mode* and is able to receive system information from the PLMN. While in idle mode, the UMTS Radio Access Network (UTRAN) has no information of the mobile, which is only identified by the non-access stratum. The mobile passes from idle mode to *connected mode* when a

Table 5.1 Transport channels for user data transmission in WCDMA

	DCH	DSCH	Common channels		
			FACH	RACH	CPCH
Uplink/downlink	Both	Downlink	Downlink	Uplink	Uplink
Code usage	According to max DCH bit rate	Code shared between users	Fixed codes per cell	Fixed codes per cell	Fixed codes per cell
Fast power control	Yes	Yes	No	No	Yes
Soft handover	Yes	No	No	No	No
Suited for	Medium/large data amounts	Medium/large data amounts and very bursty traffic	Small data amounts	Small data amounts	Small/medium data amounts

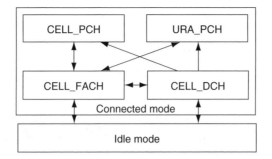

Figure 5.10   RRC connection states in WCDMA

Radio Resource Control (RRC) connection is established between the UTRAN and the mobile. While in connected mode, the mobile can be in four possible states (Figure 5.10). CELL_DCH is the only state allowed when a DCH is established to the mobile in uplink or downlink. The mobile is known on cell level according to its active set. In CELL_FACH state the mobile continuously monitors the FACH in the downlink. In CELL_FACH the mobile is assigned a default common channel in the uplink (e.g. RACH) that it can use anytime according to the access procedure for that channel. In CELL_FACH, the position of the UE is known by UTRAN on cell level according to the cell where the mobile last made a cell update. In CELL_PCH state the uplink activity is not possible, and the mobile uses discontinuous reception to monitor the paging channel. URA_PCH state is very similar to CELL_PCH, but the location of the UE is known on UTRAN Registration Area (URA) level instead of cell level.

The only two states that allow user data transmission are CELL_DCH and CELL_FACH. When a packet data transmission is to be started, UTRAN has to decide which channel should be allocated. Traffic volume reporting is used to select the channel type (common or dedicated channel). The traffic volume reporting can be done in two ways: periodical or event triggered [5]. The MAC layer receives information from the RLC layer about its Buffer Occupancy (BO).

Typically, if the BO exceeds a given threshold during a certain period of time, UTRAN selects the DCH (or DSCH) as the conveying transport channel and a capacity request is sent to the Packet Scheduling functionality. If the scheduler grants the capacity request, a DCH is established between the mobile and the UTRAN. If the BO does not exceed the aforementioned threshold, UTRAN selects a common channel for the data transfer. Figure 5.11 depicts an example of traffic volume report triggering.

Due to the delay required to establish the DCH, it is not released immediately after the RLC buffer runs empty. Instead, a DCH *release timer* may be set in order to measure the time the BO equals zero. If the timer expires, the DCH is released, and the mobile returns to CELL_FACH state.

From a service performance viewpoint, the transmission of packet data over WCDMA is shaped by the RRC connection state changes. For example, in an FTP download, the packets corresponding to the initial TCP connection establishment are typically transferred over common channels (i.e. low bit-rate transmission). Then, when the actual data are retrieved,

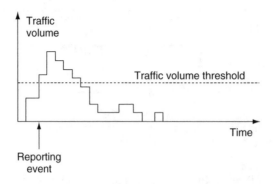

Figure 5.11   Traffic volume monitoring in WCDMA

the BO usually exceeds the traffic volume threshold, and a capacity request is sent to the Packet Scheduling functionality. The processing of the capacity request plus the establishment of the DCH incur a significant delay that has two immediate effects: (i) delays the overall file transfer and (ii) suddenly rises the Round-Trip Time (RTT) (section 5.4.1.4 – the effect of the sudden RTT increases in TCP). During, the CELL_DCH state, medium to high bit rates are typically available (depending on the overall cell load). Finally, the DCH release timer has a direct impact on the user's QoS because consecutive downloads (e.g. in HTTP browsing) may utilize the same DCH (avoiding the DCH establishment delay) if the release timer is long enough.

5.3.2.1 Sources of Throughput Degradation for DCHs in WCDMA

Applying a similar bottom-up approach from physical to IP layer as in (E)GPRS, the following factors determine the end-user throughput in WCDMA (Figure 5.12).

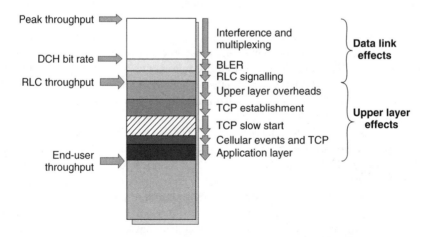

Figure 5.12   Sources of throughput degradation for DCH in WCDMA

- the peak throughput
- the resource multiplexing among users (and the effect of the interference)
- the Block Error Rate (BLER)
- the RLC signalling.

The *peak throughput* is the maximum bit rate that the physical layer can provide under ideal signal quality conditions. In WCDMA, the peak throughput of the physical layer is determined by the chip rate, the Spreading Factor (SF) of the spreading code used and the coding rate. Table 5.2 shows the downlink peak throughput for different SF configurations [4]. The minimum SF (SF=4) can provide a physical layer throughput of 936 kbps. Multi-code operation[3] can enhance the maximum available bit rate up to 2.8 Mbps.

In WCDMA, all the users under the same frequency in a given cell have to share the cell resources. The cell resources to be shared among users are the air interface capacity and the physical resources (e.g. the spreading codes of the code tree). These resources are managed by the Packet Scheduler and Resource Manager functionalities.

The Resource Manager has the functionality of handling the allocation of physical resources. The Resource Manager allocates scrambling codes in uplink and spreading codes in downlink. The allocation of scrambling codes in uplink does not represent a serious limitation due to the large set of available scrambling codes. However, the allocation of spreading codes in downlink is subject to code blocking due to the limited amount of these codes. Figure 5.13 depicts the code tree of spreading codes in WCDMA. Note that there exist N codes of SF N. Additionally, there exists a limitation in the allocation of the codes: a channel may use a certain code in the tree if no other channel uses another code located on an underlying branch of the tree. Then, the allocation of high bit rates (with low SFs) consumes a large fraction of the code tree and thereby limits the number of users that can simultaneously transmit in the downlink.

The Packet Scheduling functionality is the one in charge of dividing the available air interface capacity for non-real-time users. The scheduler typically operates periodically and on a cell-level basis. As described in the previous section, when the amount of data in the RLC buffer (uplink or downlink) exceeds the traffic volume threshold for a given period, either

Table 5.2 Maximum physical layer user throughput in downlink WCDMA

Spreading factor	Channel symbol rate (kbps)	Channel bit rate	DPDCH channel bit rate range	Maximum user throughput (~1/2 coding rate)
64	60	120	90	45 kbps
32	120	240	210	105 kbps
16	240	480	432	215 kbps
8	480	960	912	456 kbps
4	960	1920	1872	936 kbps
4, with 3 parallel codes	2880	5760	5616	2.8 Mbps

---

[3] Note that four codes of SF 4 cannot be used for DCH purposes because this allocation would consume the entire code tree. A fraction of the code tree must be reserved for common channel purposes.

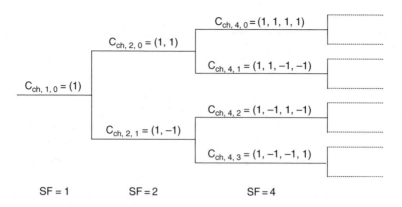

Figure 5.13   Code tree of orthogonal variable spreading factor (OVSF) codes in WCDMA

the mobile or the network sends a request for capacity to the cell-specific packet scheduler. The capacity requests are queued in a buffer until the packet scheduler serves them. Prior to the bit-rate assignment, the Packet Scheduling functionality has to measure the current cell load[4] from real-time connections (as well as load from other cells). With that measurement, and with the target load level in the cell, the scheduler can have information of the available capacity for new packet data allocations. There exist several scheduling methods to distribute the available load among the users' requesting capacity. Depending on the packet scheduling method, both the network and the user performances will vary. Three typical scheduling methods are:

1. *Fair resource*: Each user gets equal resource. This means that user location will determine the effective data rate of the connection (depending on its C/I conditions).
2. *Fair throughput*: All users will get the same throughput independent of their location. This technique distributes the throughput evenly among the users in the cell, but it is not very optimal from the system efficiency point of view because the users at the cell edge consume a large amount of resources.
3. *C/I scheduling*: Users with good C/I will have higher priority than users with low C/I. This method obtains the highest system performance at the expense of the lack of fairness because the users with poor radio propagation conditions get lower amount of resources.

In micro-cell deployments, where the propagation of the radio signal suffers low time dispersion, the effect of the interference is reduced and the air interface capacity increases significantly. In such case, if the traffic is very bursty with long inactivity periods, the overall downlink cell capacity may be limited due to the lack of spreading codes. In general, for macro-cell scenarios, it can be said that the air interface capacity imposes a

---

[4] Typical metrics of air interface load are the total wideband received power in uplink, and the total cell transmission power in downlink.

tougher limit on the overall cell capacity than the physical resources (for both uplink and downlink).

It is relevant to note that bit-rate capabilities of the WCDMA terminal are limited (e.g. 384 or 512 kbps in legacy terminals), and the packet scheduler must take these capabilities into consideration in the DCH bit-rate allocation process.

Figure 5.14 depicts the relation between the spectral efficiency and the average user throughput in WCDMA and other technologies. The figure implicitly assumes no physical resource limitation. For high user throughput, the number of users in the cell is low and the spectral efficiency decreases; and vice versa, for large number of users in the cell and high spectral efficiency, the average user throughput is reduced. One of the interesting conclusions from Figure 5.14 is that WCDMA allows much higher user throughput than (E)GPRS/CDMA2000 due to the larger bandwidth.

The link level throughput that can be achieved by a certain user at a given time is a function of the allocated resources (defined by the Packet Scheduling functionality), the interference environment (e.g. users at the cell edge can obtain less throughput) and the quality (BLER) target. In this respect, when the BLER target is too restrictive, more power is needed, which lowers the system throughput (or capacity) due to the fact that power is a limited resource. On the other hand, if the BLER target is too high, many retransmissions are needed, which also consume resources. Taking into consideration that retransmissions degrade the end-user performance, adjusting the proper BLER target involves an implicit trade-off between quality and system capacity.

In acknowledged mode, the peers of the RLC layer exchange control information to ensure the error-free delivery of the data. However, the effect of this signalling overhead is expected to be minor, especially for high bit-rate allocations.

As described in the (E)GPRS case, delay variations or gaps in the transmission may produce some degradation in upper layers performance. In WCDMA, such delay variations may be due to retransmissions at RLC level to achieve the target reliability, RRC state

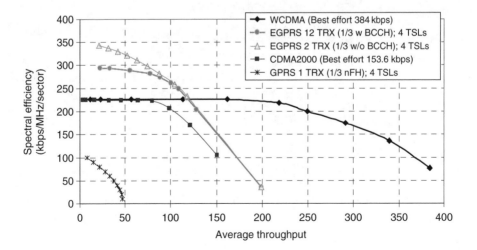

Figure 5.14 Spectral efficiency vs average user throughput in various technologies

change, loss of radio coverage, transmission stopped or downgraded because of higher priority traffic (depending on the scheduling method) or due to hard-handovers.[5]

5.4 Transport and Application Layer Effects

Before going into the details of transport and application layers itself, another basic concept must be considered in the performance degradation analysis along layers. It is referred to the overhead introduced by upper layers, which depends on the type of service under analysis.

An example of protocols overhead introduced in TCP-based services is shown in Figure 5.15. The impact of header overhead is smaller as IP packets length increase. Therefore, when using TCP-based services it is generally recommended to use big TCP segments in data transactions (e.g. 1460 bytes in Ethernet as described in section 3.3.1) in order to enhance application throughput and end-user performance. However, the optimum IP packet length for non-TCP-based services depends on the traffic pattern and requirements.

In some cases, some header compression schemes could be applied to minimize such overhead, although it might not perform well if too many errors or disordering occur between the two compressing nodes. In addition, it generally implies higher processing power and memory consumption in the terminal and server (see section 8.4 for further details on compression techniques). The following subsections will concentrate on the impact of transport and application layers on the overall performance.

5.4.1 TCP Performance

Wireless network provides packet data transfer capabilities that are used by transport and application protocols for exchange of information. At transport level, there are several protocols that can be used, depending on the service. For error-free message delivery, TCP is

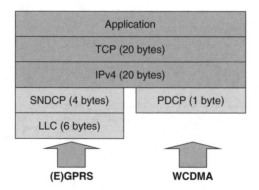

Figure 5.15   Protocol overheads for TCP-based services

---

[5] Examples of hard-handovers in WCDMA are: inter-system handovers (between two different technologies) and inter-frequency handovers.

commonly used. Actually, the majority of Internet traffic is implemented over TCP because it is a stable and probably the most thoroughly tested transport protocol of its kind.[6] However the throughput performance of TCP is very sensitive to the characteristics of the communication pipe. Factors such as latency, throughput and packet losses have great impact on the performance of the TCP layer, and interact very much with TCP, its congestion control and retransmission algorithms. One reason is that TCP protocol was originally designed for wired networks, where the main cause of packet losses is the network congestion (packet losses due to data corruption is typically much lower than 1%). However, the inherent characteristics of wireless networks significantly differ from the assumptions above, as explained in section 3.3. In the following section, the effects of the communication over wireless networks on TCP are analysed.

5.4.1.1 Wireless Network Characteristics Affecting TCP

As already pointed out in previous sections, wireless networks behaviour differs considerably from wired networks. Characteristics of the link layer of wireless networks (i.e. 2.5G/3G) have significant effects on TCP and application layers performance. Big latencies, variable data rates, asymmetry, delay spikes, data loss or bandwidth oscillations are some of the main effects that can affect the performance of TCP. For this reason, it is important to understand how these different effects interact with the protocol, in order to find out possible improvements or parameter optimizations that could help to reduce any degradation [8]. Obviously, the radio technology under consideration is a differential factor. However, every network can be characterized by few basic metrics, and these can be directly related to an impact of the transport and application protocols.

This section is trying to extract these general characteristics of wireless environments, moving later to some particular effects of different wireless technologies.

Latency: TCP slow start mechanism jeopardizes the end-to-end performance to a large extent when RTT is relatively high, as it is the case of cellular networks. This high delay may be caused due to long transmission and processing delays in the RAN and Core Network (CN). A typical RTT varies between few hundreds of milliseconds and 1–2 seconds, depending on the radio technology and network architecture. Large latencies lead to larger TCP connection establishment delays and slower recoveries from TCP slow-start mechanism.

Packet Losses: Packet losses in wired networks are mostly due to buffer overflows under high load conditions. On the other hand, wireless networks usually have errors in the radio link (Bit Error Rate or BER). Although correction methods (e.g. ARQ and FEC techniques) can minimize the undetected errors, others cannot be solved, leading to a residual BLER. Moreover, blocks could be also lost due to mobility procedures which require dynamically moving the information from one base station to another. Therefore, there are different reasons for packet losses in wireless networks than congestion, as it is assumed by TCP. As described in Chapter 3, TCP reacts against packet losses by limiting its transmission rate, as a way to control the network congestion. This is achieved by reducing the TCP congestion window (cwnd) and slow start threshold (ssthresh) to the half.

---

[6] According to measurements [22], around 95% of the overall data traffic crossed by a middle Internet backbone is TCP traffic.

Delay Spikes: A delay spike is a sudden increase in the latency of the link. It is possible that a mobile terminal suffers sudden delays because of cellular specific events, such as:

- Bad quality in the link that increments the delay due to RLC retransmissions; also loss of radio coverage, for example, while driving into a tunnel or while within an elevator.
- Transmission stopped when changing from one cell to another (as described in section 5.3). During a cell change, the mobile terminal and the new base station must exchange messages and perform some other time-consuming actions before data can be transmitted into a new cell. Many wide-area wireless networks try to provide seamless mobility by internally re-routing packets from the old to the new base station which may cause extra delay.
- Transmission downgrading or blocking by high-priority traffic may occur when a new CS call or higher-priority data temporarily pre-empts the radio channel. This happens because most current terminals are not able to handle a voice call and a data connection simultaneously, and suspend the data connection while speech call is ongoing. Additionally, a scheduler in the radio network can suspend a low-priority data transfer to give the radio channel to higher-priority users.

Sudden increase in transmission delay can cause spurious TCP timeouts, triggering unnecessary retransmissions and producing a new slow start phase [1].

Data Rates: The main reason for moving forward from 2G to 3G, in addition to increases on voice capacity, are the higher bit rates for data services. Typical rates for 2G systems are about 10–20 kbps in uplink and 10–40 kbps in downlink, and up to four times more for 2.5G, while for 3G systems it is expected to have bit rates around 64 kbps in uplink and 384 kbps in downlink. The resulting BDP in cellular networks is around 1–5 kB for 2G, 20–25 kB for 2.5G, and 8–50 kB for 3G. According to these values of BDP, the links can be classified in long thin networks (high delay, low throughput), and long fat networks (high delay, high throughput). In both cases it is very important to keep a long enough transmission window in order to obtain a good performance. Data rates are also quite variable due to effects from other users and from mobility. The number of users connected to one cell can reduce or increase the available bandwidth, and, furthermore, increasing the distance from the base station decreases the link bandwidth due to reduced link quality. Finally, by simply moving into another cell, the user can suffer sudden changes in the available bandwidth and interruptions of the transmission during few milliseconds, or even seconds. Note that even if TCP tries to adapt to the link bit rate, sudden changes in this capacity may make TCP to underuse the link capacity, use a too-high transmission window, or even perform unnecessary retransmissions due to retransmission timer expiration (i.e. spurious transmissions [1]).

Asymmetry: 2.5G/3G systems typically have asymmetric uplink and downlink data rates. The uplink data rate is limited by battery power consumption and complexity limitations of mobile terminals. However, the asymmetry does not exceed 3–6 times, and can be tolerated by TCP without the need for techniques like ACK congestion control or ACK filtering.

A summary of the main differences between wired and wireless networks, as well as the effect on TCP is shown in Table 5.3.

Table 5.3 Networks characteristics and impact of TCP

Indicator	Wired	Wireless	Effect on TCP
Latency	Very low latencies in general	High latencies mostly due to transmission delays in the RAN or CN	Larger TCP connection establishments and lower recoveries from slow start
Packet loss	Low packet losses (only due to buffer overflows)	Can be attributed to corruption (BLER) instead of congestion	Three consecutive similar ACKs triggers TCP retransmission, CWND and threshold to the half (Chapter 3)
Delay spike	Low delay spikes in general	Important delay spikes due to different wireless procedures	Delay spikes can cause spurious TCP timeouts, leading to unnecessary retransmissions, a decrease in the CWND to one segment
Data rates	High data rates	Low and dynamic data rates	Underutilization of the radio link and TCP timeouts

Another important TCP characteristic which has a clear impact on the performance is the fact of sending ACKs for every one or two segments, which implies that the radio interface is used in both directions. In principle, this aspect might not imply an important drawback since the data rates used for this purpose is not so high in the reverse direction. However, ACK traffic in the reverse link should not be limited by bit rate or delay if we want to ensure an optimum TCP performance. The speed at which the TCP transmitter increases its bit rate and how congestion is detected depend on delay suffered by ACKs and the probability that they are lost. For instance, in any wireless network, the following issues need to be taken into account.

- ACKs can be delayed or even get lost in the mobile network.
- TCP bases its sending rate on the ACKs arrival rate (congestion control limitation). This ACK traffic is generally bursty traffic, and if the wireless network does not consider this effect and assigns radio resources only for the transmission of each packet, the arrival rate is usually increased by establishment delay of the radio channels.
- ACKs produce some signalling at lower levels, which consume resources in the other direction (i.e. (E)GPRS, as explained in section 5.3.1).

All these factors must be considered and optimized in order to achieve an optimum TCP performance. In Chapter 8, different TCP optimization techniques are also described.

5.4.1.2 The Bandwidth Delay Product

In the ideal steady state of a TCP connection, the sender injects segments into the data pipe at the same pace as the receiver removes them from the network. In such state, the number of ACKs making the return trip equals the number of segments in the pipe. Figure 5.16 shows the transmission of TCP segments and ACKs in ideal steady state.

The BDP is a figure of merit representing the amount of outstanding data required to achieving maximum throughput in the ideal steady state of a TCP connection. The BDP can be computed by the following formula.

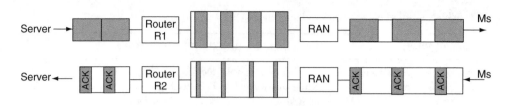

Figure 5.16   TCP transmission in ideal steady state

$$BDP(bits) = Bandwidth(bps) \cdot Delay(s)$$

where Bandwidth represents the bandwidth of the most limiting link in the end-to-end connection (most likely the air interface) and Delay is the RTT of the cellular network. The RTT value employed for the BDP calculation is assumed to be measured with two implicit conditions: (i) the TCP segment measuring the RTT does not experience a large queuing delay at the buffers of the network nodes caused by segments belonging to the same TCP connection and (ii) the RTT includes the transmission delay caused by the size of the TCP segment.[7] Some typical values of the BDP in different radio technologies are shown in Table 5.4.

Ideally, the congestion window of the TCP connection should be slightly higher than the BDP because lower windows cause the limiting link (radio in this case) to be underused. The latter situation represents a disadvantage for the end-user who experiences a connection throughput below the possibilities of the radio link. But it also has disadvantages for the RAN because the physical resources are assigned to the users without fully exploiting them. For example, in WCDMA the allocation of high bit rates requires the assignment of large amount of channelization code resources. If those code resources are not fully utilized, the RAN may suffer from code shortage, thereby restricting the cell capacity [5].

On the other hand, congestion windows larger than the BDP imply more data than what the limiting link can handle, which causes the data excess to be cumulated in the buffer of the bottleneck link.

$$CWND(bits) = BDP(bits) + B(bits)$$

where CWND represents the size of the congestion window and B indicates the average amount of queued data in the buffer of the bottleneck link. Large amounts of data excess

Table 5.4   Example of BDP

Technology	Bandwidth (in kbps)	Delay (RTT)[a] (in ms)	BDP (in kB)
GPRS	30	1200	4.5
EGPRS	150	800	15.0
WCDMA	300	300	11.25
CDMA2000-1X	100	450	5.625

[a]RTT assuming a 1460 bytes in DL and 40 bytes in uplink. This is typical RTT affecting during TCP slow start.

---

[7] For example, the transmission delay at the radio link significantly differs from 40-byte to 1500-byte packets.

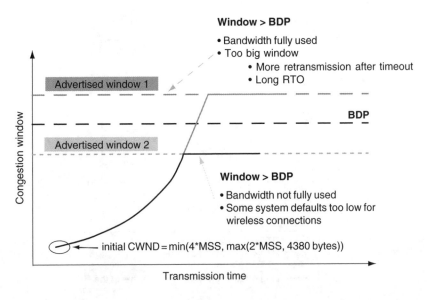

Figure 5.17   Advertised window and bandwidth delay product

queued in the buffer of the radio link may cause buffer overflow, with subsequent segment losses.

The slow start algorithm increases the congestion window size aggressively in order to reach the BDP as swiftly as possible. The objective of this aggressive increase is to minimize the period when the limiting link is underused. Since the BDP indicator measures the amount of outstanding data that would 'fill the pipe' (i.e. the available capacity of the cellular link), it is really important that the advertised window (awnd) in the receiver side should be larger than BDP in order not to limit the congestion window (Figure 5.17).

5.4.1.3 Influence of the RTT on the Service Performance

The RTT constitutes one of the most important sources of potential performance degradation for TCP-based services.

Since each TCP connection establishment takes approximately 1.5*RTT (section 3.3), both the number of TCP connections and the RTT values should be minimized as much as possible in wireless environments. During slow start, the RTT value also plays an important role, given that TCP congestion window can be increased only on a per RTT basis. The duration of the slow start will be inversely proportional to the RTT and maximum achievable bandwidth. The longer the RTT, the slower will be the congestion window growth (as the ACKs need more time to reach the sender); on the other hand, the higher the bandwidth, more the number of cycles are required to reach the maximum capacity. Additionally, long delays affect the retransmission algorithms by delaying the response time for recovering transmission errors.

The relative impact of the slow start on the TCP throughput performance depends on the size of the file to be transferred. As slow start can be characterized as a constant delay in the beginning of the transmission, the effect on throughput degradation increases with smaller file sizes. This is due to higher relative time the slow start algorithm takes in the whole TCP transfer. Figure 5.18 depicts an example of the FTP throughput in EGPRS and WCDMA for

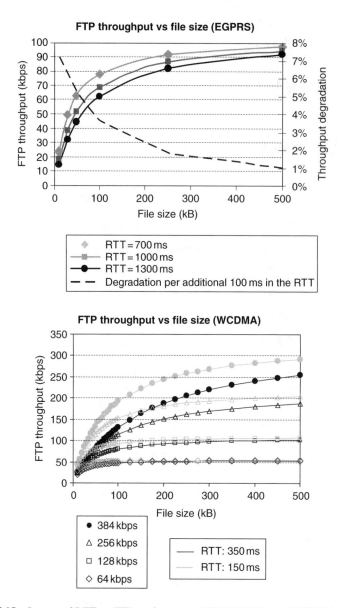

Figure 5.18   Impact of RTT on FTP performance (EGPRS 3TSL and WCDMA example)

typical values of the RTT in both technologies. As mentioned in the previous section, the RTT in Figure 5.18 is assumed to be measured when the queuing delay of TCP segments of the same connection is negligible. The WCDMA further assumes the continuous allocation of a DCH for the whole FTP retrieval.

In this example, an EDGE-capable terminal suffers up to 7% of throughput degradation per additional 100 ms in the RTT in the worst case. However, the effect of the RTT is more noticeable for small file sizes due to slow start. Moreover, the impact of the RTT on the TCP performance also depends on the bit-rate capabilities (e.g. EGPRS further affected compared to GPRS, or different bit-rate allocations in WCDMA). During slow start, connections with higher potential available bandwidth will be wasting more resources than other connections with lower available bandwidth. For that reason, the throughput degradation as RTT increases is higher for those connections with higher available bandwidth. In any case, the optimization of the RTT in the network is a key aspect to enhance the overall service performance.

As mentioned in section 5.4.1.2, the TCP advertised window may limit congestion window to a value below the BDP, thus reducing the TCP throughput below the bottleneck link capacity. Figure 5.19 plots the FTP throughput as a function of the RTT in WCDMA (DCH) for a 250-kB file. The results show that the needs for TCP receiver's buffering capabilities completely vary depending on both the available bandwidth and the RTT. It can be seen that there exists no throughput performance difference between the tested awnds for 64-kbps DCHs and for realistic round-trip times. The 8-kB receiver's window only limits the throughput for 128-kbps channels for RTTs larger than 300–400 ms. Larger windows are definitely required for higher bit rates.

It is clear from Figure 5.19 that the larger the TCP receiver's window, the wider is the range of bit rates, and RTTs can enjoy full radio bandwidth utilization. However, having very large receiver's window has a relevant disadvantage besides the cost of the memory in the mobile terminal. The TCP transmitter does not have information of the actual BDP, so the congestion will grow over the BDP until there is an error or it reaches the awnd. Hence, as commented in the previous section, the data excess over the BDP will be cumulated in the buffer of some network element before the limiting link (typically SGSN). With a relative segment loss free connection, if the RTT is low and the allocated channel bit rate is low too, the amount of data to be queued may cause such buffer to overflow. Note that the buffering capabilities (SGSN in (E)GPRS and RNC in WCDMA) must be sufficient to store all the information of all the connections. The buffer overflow is a very much undesired situation because it will drop

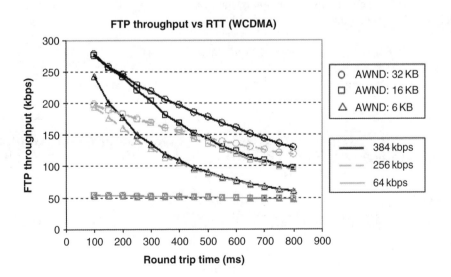

Figure 5.19   FTP throughput for various DCH bit rates and receiver's buffer capabilities in WCDMA

multiple TCP segments that will trigger a retransmission timer expiration with its subsequent negative effects on the congestion window and ultimately in the user throughput.

It is very important to analyse the dynamics of the Round-Trip Time since a sudden increase of the RTT above the Retransmission Timeout (RTO) may cause timer expiration. At the beginning of a TCP connection, RTT values are typically low and as TCP window grows, higher buffer occupancy leads to longer delays. For good radio conditions, the congestion window will increase until reaching the advertised window in a steady state.

In steady state, the *total Round-Trip Time* (RTT*) can be computed by adding the queuing delay at the bottleneck link:

$$RTT^* = RTT + B/BW = (1/BW)(RTT \cdot BW + B) = CWND/BW$$

where
    RTT* represents the total Round-Trip Time measured in seconds
    RTT is the Round-Trip Time employed for the BDP calculation[8] (i.e. excluding the queuing delay)
    B indicates the average amount of queued data in the buffer of the bottleneck link
    BW is the bottleneck bandwidth and
    CWND represents the TCP congestion window size.

For example, Figure 5.20 depicts the RTT* experienced by the segments in a TCP connection over a 64-kbps DCH in WCDMA. The receiver terminal is assumed to be equipped with an 8-kB TCP receiver buffer, the BLER target is set to 10%, and the RTT equals 300 ms. In steady state, it can be seen that the average RTT* approximates to:

$$CWND/BW = 8\,kB/(64\,kbps \cdot 0.9) \approx 1.14 \text{ seconds}$$

Additionally, under the same conditions, the average amount of queued data at the bottleneck link can be computed as:

$$B(bits) = CWND(bits) - RTT \cdot BW$$

For the above example, the amount of queued data $B = 8\,kB - (0.3\,s \cdot 64\,kbps \cdot 0.9) \approx 5.9\,kB$

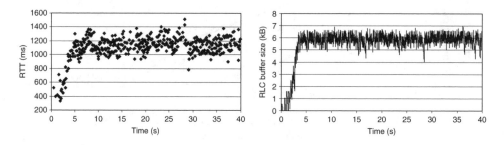

Figure 5.20   Example of WCDMA RTT evolution along the time

---

[8] See the definition of the RTT for the BDP calculation in section 5.4.1.2.

5.4.1.4 The RTO and the Spurious Timeouts

As explained in Chapter 3, TCP keeps a retransmission timer to detect packet losses due to congestion somewhere in the network between the source and the destination. TCP triggers congestion control actions if a packet is not received before the retransmission timer reaches the RTO [6]. Ideally, the RTO should be able to closely follow the RTT along the connection (see also section 3.3.1 for further details on RTO computation). The RTO should be low enough to swiftly react to the packet losses caused by congestion in the network. On the other hand, the RTO should permit fluctuations of the RTT in order not to cause unnecessary timeouts.

A spurious timeout occurs when, despite there exists no packet loss, an unnecessary timeout is triggered because the RTT suddenly increases exceeding the RTO. The spurious timer expirations are very unfortunate because the sender must reduce the slow-start threshold, i.e. the long-term offered load, and it is forced to execute slow start to retransmit all the segments that had already been transmitted but not acknowledged (which may not even require retransmission). Besides, the double transmission of the packets can trigger duplicated ACKs, which may activate the fast retransmission. The ultimate result is that the congestion window, and therefore the user throughput is significantly decreased [1].

As described in section 5.4.1.1, delay spikes and data-rate variation constitute the main reasons to trigger TCP timeouts under wireless. Depending on the technology, such events may be caused by different reasons (cell reselections in (E)GPRS; sudden bit-rate reductions or DCH establishment delays in WCDMA), although the vulnerability of TCP comes from that effect, not from the cause.

If a wireless event happens during the first part of the connection (transitory state), there is higher probability of a timeout than in the case of steady state, where RTT and RTO keep higher values, adapted to the transmission window size. This is due to the higher and quick variation of the RTT at the beginning of the connection, and additionally the lower receiving rate of segments in transitory state makes the retransmission timer to reach higher values. Moreover, during the early stages of the connection, the congestion window is typically small, and therefore halving the slow start threshold down due to the time out sets the starting point of congestion avoidance very low, significantly limiting the speed at which congestion window can be incremented, and thereby, increasing the time during which congestion window is smaller than the BDP (throughput degradation).

Figure 5.21 shows two examples of RTT and RTO behaviour under EGPRS networks with different outage time durations when performing cell reselections [7]. An EDGE capable mobile moving at 50 km/h with 3 + 1 TSL capability has been traced while performing an FTP download. Advanced TCP options (TimeStamp and SACK) are enabled. These features are advanced TCP options recommended for cellular networks [8] and explained in section 8.3.1. An advertised window of $14 \cdot$ MSS was set in the receiver, according to the BDP computation (930 ms as delay, estimated from RTT of ping packets and a maximum achievable throughput of 165 kbps).

It is important to understand that TCP, without TimeStamp, estimates the instantaneous RTT once per TCP window. This aspect is mainly harmful in networks with high BDP because window size is usually high and RTT is estimated less frequently. Under these conditions, RTO is not adapting as quickly and accurate as desired. However, TCP TimeStamp option allows estimating the RTT per segment, providing a better performance in general.

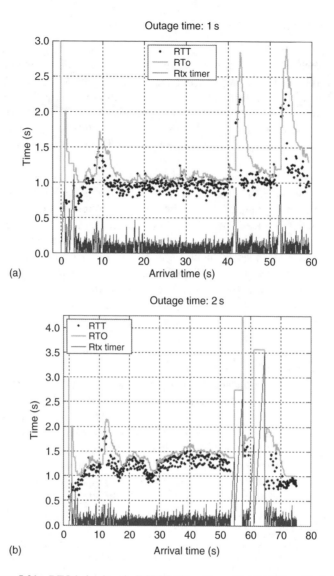

Figure 5.21   RTO behaviour in EGPRS (TimeStamp and SACK enabled) [7]

In the Figure 5.21, every time a new ACK arrives to the server, the retransmission timer is reset, and after that it starts increasing. Whenever the timer reaches the RTO value, a retransmission takes place. In this example, 1-second outage time (Figure 5.21a) is not high enough to trigger a retransmission, since RTO is able to adapt to such variation. However, 2-second outage time (Figure 5.21b) leads to timer expiration.

However, same tests without TCP TimeStamp and SACK enabled have led to a higher probability of retransmissions, as shown in Figure 5.22 (1-second outage time). This is due to the poorer accuracy in the RTT estimation by the sender (only once per window size) and consequently the less accurate adaptation of RTO.

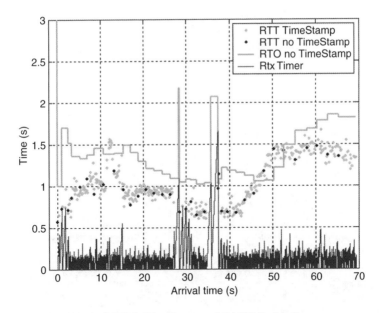

Figure 5.22 RTO behaviour in EGPRS (TimeStamp and SACK disabled)

In general, TimeStamp and SACK features improve the performance under non-ideal conditions, i.e. where there are numerous TCP retransmissions.

5.4.1.5 The Effect of Segment Losses

Previous sections of this chapter have assumed an error free TCP connection. Under those conditions, the main parameters that determine the TCP behaviour (such as the BDP, receiver buffer size, RTO, etc.) have been described. However, the appearance of segment losses, either single random drops or multiple packet losses caused by network congestion, triggers the flow-control mechanisms that completely modify the overall connection performance. In wired networks, packet losses are typically due to buffer overflow, while the losses caused by damage are very small (much below 1%). TCP has been designed with this implicit assumption. However, wireless networks suffer from link-level errors of the radio channel. Although there exist various methods to cope with link-level errors (Forward Error Correction, ARQ mechanisms, etc.), the wireless packet losses cannot be totally restored. This section will analyse the effect of such segment losses, assuming that these losses occur with a certain probability and are fully independent of each other.

Let us briefly describe the events that take place when a single segment loss occurs. Figure 5.23 depicts an example of the RLC buffer size evolution during a file download in such a situation. When the packet loss occurs, the receiver sends a duplicate ACK for every out-of-sequence data segment. The reception of three duplicated ACKs at the server triggers the fast retransmission algorithm. After the lost segment is retransmitted, the fast recovery algorithm may send a few segments if permitted by the congestion window, and when new ACKs arrive, further amount of new data can be delivered. Later, the congestion window evolves linearly as dictated by the congestion avoidance algorithm.

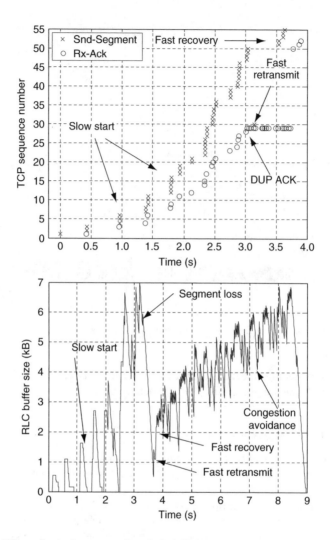

Figure 5.23 Effect of a single segment loss during file download on a 64-kbps DCH in WCDMA

If the congestion window is large enough when the random loss takes place, the retransmission and the process of halving the window may fortunately represent a minor source of throughput degradation. However, if the congestion window is small when the packet loss occurs, the long-term throughput will be significantly degraded due to the value of the slow-start threshold after the fast retransmission and recovery process.[9]

When the errors occur, the frequency of the random losses and the actual value of the congestion window are key factors in the resulting performance. If the fast retransmission and fast-recovery algorithms are triggered too frequently, the congestion window is halved too

---

[9] The slow start threshold is set to half the congestion window when three duplicated ACKs are received.

many times, and the congestion avoidance will not have enough cycles to raise the congestion window over the BDP between consecutive losses. Moreover, if more than one packet (burst of packets) are lost within the same congestion window, it will be difficult to carry out all the retransmissions and receive their corresponding ACKs before the RTO timer expires. This performance problem of TCP Reno when multiple packets are dropped from a window of data is well known [9].

Furthermore, during the initial slow start, if a segment loss occurs when the congestion window is grown-up, but the RTO has still not adapted enough to the window growth, then a timeout is likely to happen. Let us briefly explain the triggering of the timeout. When the segment loss takes place, the fast retransmission algorithm will carry out the retransmission of the missing segment. The retransmitted segment may experience an overall RTT larger than the RTO (due to the queuing at bottleneck link) and trigger the timeout.[10] Unluckily, if the random loss probability is high enough, this event is not so rare. This belongs to the undesired spurious timeouts, i.e. a timeout that could have been avoided if the sender had waited 'long enough'.

Figure 5.24 depicts the FTP throughput as a function of the random loss probability for various DCH bit rates and RTTs. The results show that the degradation of the TCP throughput

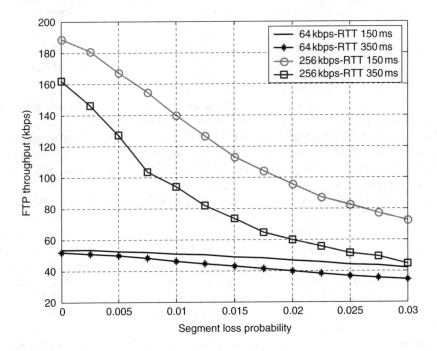

Figure 5.24   Effect of a single segment loss during file download on a 128-kbps DCH in WCDMA

---

[10] We have assumed a retransmission timer per TCP window, which is reset whenever an ACK that acknowledges new data is received, and it is turned off when all the outstanding data has been acknowledged [9].

caused by segment losses is significantly higher for high bit rates than for low ones. The main reasons are: (i) the segment transmission gap caused by either fast retransmissions or timeouts causes larger deterioration for higher bit rates and (ii) as commented above, high bit rates require larger BDPs to fully utilize the radio link.

Lakshman [10] illustrated that the effect of the random losses significantly degrades the TCP throughput if the packet loss probability exceeds a given threshold. This threshold is computed as the inverse of the square of the BDP (expressed in segments). For example, for a 64-kbps channel with a 300 ms of RTT, the probability threshold is equal to 4%, whereas for the same RTT and for a 256-kbps channel the threshold equals 0.34%.

Wireless networks (such as GERAN and UTRAN) introduce an RLC-acknowledged mode to recover the block erasures caused by the link-level errors of the radio channel. The acknowledged mode of the RLC layer aims at providing an error-free data-transmission service by making use of an ARQ mechanism that retransmits the erroneously decoded Packet Data Units (PDUs). By increasing the maximum number of RLC retransmissions (prior to PDU discarding), the packet loss rate can be reduced. However, this reliability benefit is at the expense of a variable link-level transmission delay, which can potentially lead to TCP timeouts if the retransmission timer cannot cope with the RTT peaks caused by a large number of RLC retransmissions. In general, it can be concluded that the more reliable the RLC layer is, the more benefits are obtained in terms of TCP throughput because the end-to-end TCP recovery mechanisms have negative long-term effects on the congestion window [11].

5.4.2 UDP Performance

UDP protocol does not introduce so many drawbacks and concerns over wireless due to the lack of retransmissions, flow control or congestion control mechanisms.

This protocol is much simpler than TCP, which may be good for some applications (mainly those with real-time requirements), since it allows an easier control of the delay and bit rate being provided and exploits maximum throughput available. However, this lack of control may lead to congestion.

Since UDP is commonly used by real-time applications, UDP traffic is characterized by the codec negotiated for the transmission. Optimum UDP datagram size depends on the type of application, although generally real-time services use small or medium sizes. For instance, voice packets in a push-to-talk service are around 160 bytes in length. On one hand, potential packet losses, which are not retransmitted at transport layer, only implies the loss of 160 bytes of information; on the other hand, small packet losses are very much affected by the upper layers overhead. That is why header compression techniques like Robust Header Compression (ROHC) are a key issue to optimize the performance (further details on compression methods can be found in section 8.4).

At the time of writing, new transport protocols were being developed to solve the drawbacks of UDP and TCP traffic over wireless. One of the most important ones is known as Datagram Congestion Control Protocol (DCCP), which tries to converge the philosophy of a datagram protocol like UDP with enhanced and configurable congestion control (see section 8.3 for more details on DCCP).

5.4.3 Application Layer Effects

Application layer is also introducing degradation effects from a delay perspective. Since this layer is responsible, among other tasks, for establishing sessions, processing application content or handling the required transport connections, the delays associated to aforementioned procedures must be considered in the service performance analysis.

Starting from the application signalling effect, some services (mainly those with real-time requirements) utilize application protocols to establish a session with a remote end before the data itself is transmitted. During the session establishment, both parties may negotiate some parameters related to the data transfer (e.g. codecs, port numbers, etc.). Examples of those services are:

- *SIP-based real-time services like VoIP or videoconference*: Before the real-time data is being carried between peers, they need to establish an SIP session according to [12][13], where both peers negotiate the end-to-end QoS parameters (e.g. by using session description format SDP [14]).
- *Streaming services*: They normally use the Real-Time Streaming Protocol (RTSP) to establish sessions and control the delivery of real-time streaming data [15] [16] (Figure 5.25).

The session set-up delay depends on different factors like:

- *Network delay*: Processing delays in the different network elements affects considerably the end-to-end negotiation. Especially, when IMS takes part in the negotiation, overall network delay is increased during this phase.
- *Bit-rate capabilities*: Although application packet sizes during session establishment phase are usually not too big (between 100 and 1000 bytes), radio capability may also affect to the total delay mainly when no compression method is applied.
- *Compression methods*: Several methods are being developed to compress as maximum as possible the application messages between peers. For instance, Signalling Compression (SigComp) [17] is a solution for compressing messages generated by application protocols

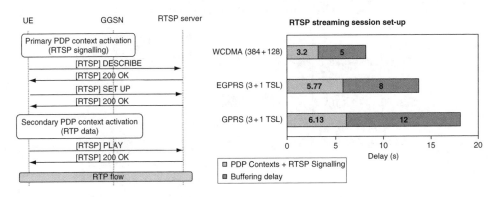

Figure 5.25   RTSP-based streaming session establishment

such as SIP or RTSP. Some of these methods may use static or dynamic dictionaries to optimize the compression ratio.

- *PDP Context usage*: The higher the number of PDP contexts required by the application, the higher the overall session setup delay. However, the use of different PDP Contexts for different flows will provide a better QoS during the service delivery.
- *Application buffering delay*: Real-time services may use buffers at application layer, which are used to compensate possible delay variations in the network. From the end-user perspective, this delay can be considered as part of the session establishment since the application does not start reproducing the media until the buffer is full. Especially in streaming services, the buffer length may be from 1 to 10 seconds, in order to optimize the service continuity. This buffer is normally a configurable parameter that must be optimized depending on the network characteristics. Buffer lengths in conversational services are generally very small (not higher than 500 ms) since the interactivity requirements are much stricter than in streaming services.

Non-real-time services do not generally require any session set-up beforehand, although application messages also play an important role in the end-user performance (e.g. Web page download cannot start till HTTP requests are received in the server).

Under wireless environments, it is also important to make an efficient use of the TCP connections by the application layer, since TCP connection delays and slow starts are an important source of performance degradation. In this respect, Web service has been one of the most relevant areas of research since the old HTTP version (HTTP 1.0 [18]) has been proved to be non-optimum over wireless. However, HTTP 1.1 [19] includes some features like *Persistent TCP connection* (mandatory) or *Pipelining* (optional) that enhances the TCP usage and HTTP performance in general. Persistent connection and Pipelining features allow reducing the number of TCP connections to be established and certain parallelism when requesting the different objects, respectively (as was described in section 3.6).

Web browsing performance results for different HTTP versions and radio technologies are shown in Figure 5.26. HTTP 1.1 may achieve performance gain around 10–20% in terms of

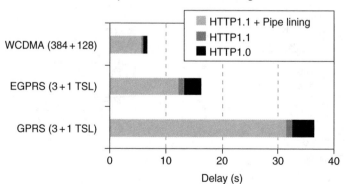

Figure 5.26   HTTP performance (100-kB Web page: 20-kB text + 80-kB objects)

total downloading time. Such a gain is mainly due to the fewer number of TCP connections that need to be established (i.e. higher reusability) and the gain depends not only on the radio capability and RTT, but also on the number of objects included in the Web page.

5.5 Impact of Network Dimensioning in the Service Performance

Cellular networks were initially designed only for voice services utilizing dedicated resources. This approach is suitable for real-time services in order to fulfil the required QoS, i.e. to provide a constant bit rate as well as a negligible and fixed delay. This approach has also the advantage of simple dimensioning rules, e.g. by using the Erlang B formula. This formula relates the amount of resources and traffic load with the desired blocking probability, i.e. probability of lack of resources, by assuming mean holding time exponentially distributed and a *Poisson* process for call attempts, i.e. no relationship between call arrivals.

However, the introduction of data services makes impossible to dimension the wireless resources by using this simple formula. That approach is not valid due to three main reasons.

1. *Traffic characterization*: Voice traffic is mainly characterized by the mean holding time and call arrivals rate. However, there are many types of data services with very different traffic characteristics. For example, FTP service generates a more constant traffic whereas Web browsing is more bursty due to interactivity with the user, protocol procedures for downloading the different objects within a Web page, etc.
2. *Quality of service (QoS)*: Different data services will demand different QoS requirements (mainly bit rate and delay) to the wireless network, which shall have mechanisms to fulfil those requirements. For instance, conversational services will require low and constant delay whereas interactive traffic will have less restrictive delay requirements. The different treatment of the traffic flows by the network shall be also considered in the dimensioning of resources.
3. *Resources allocation*: The burstiness nature of data services imposes that resources shall be shared between different services and users in order to optimize their usage.

To summarize, different traffic characterization of data services makes it impossible to continue using the previous dimensioning rules utilized for voice services, and hence network dimensioning shall consider services and users multiplexing as well as different quality requirements.

In order to guarantee that QoS requirements are fulfilled, all network elements and interfaces shall be correctly dimensioned and configured. For instance, real-time services require that all network interfaces guarantee a determined user throughput. If any of the interfaces cannot fulfil this requirement, the end-to-end user throughput will not be supported. In the same way, delays in different network elements and interfaces are cumulative, meaning that end-user delay will be the sum of all of them. Therefore, each individual delay shall be below a certain limit in order to fulfil the end-to-end delay requirement.

The dimensioning process varies for the network elements and interfaces of RAN and CN domains due to their different characteristics. RAN is characterized by a low and non-constant bit rate as well as variable delays since they depend highly on radio link conditions. Therefore, a proper dimensioning and configuration will have a big impact on the end-user perception, e.g. throughput in radio interface should be taken into account for admission. In addition to this, since radio resources are shared between several users, the scheduler policies will determine how to fulfil the QoS requirements for each service.

However, CN is characterized by high bit rates and low delays with small deviation, in general, being the interaction with the RAN part, and congestion control mechanisms the cause for relatively higher delays in this domain. Scheduling policies in the CN will have an impact on the end-user performance but less noticeable than in RAN due to higher throughput and more constant network conditions. Since CN conditions are not affected by radio interface, which is a high variable environment, its dimensioning follows simpler rules, typically based on traffic load and other limiting factors (e.g. certain number of traffic channels (TCHs), sectors, etc. supported by the network element).

5.5.1 Dimensioning Example for (E)GPRS Services

The following subsections will show the main issues in GSM & (E)GPRS networks that will impact on the user experience. They can be grouped on issues related to the radio interface (radio link conditions, multiplexing effects) and related to other network elements and interface, i.e. near the core network (A-bis interface, BSC/PCU, SGSN).

The content of this section is intended to show trendings and how these factors affect the end-user performance. However, the different figures and numbers are shown as examples and should not be understood as trustable values.

5.5.1.1 Radio Interface Dimensioning

Radio interface is characterized by being a very variant environment from a link level perspective. However, user performance is affected not only by radio link conditions but also by the RRM. (E)GPRS RRM allows multiplexing the radio channels by several connections (up to certain limit). The following formula takes into account the different effects that impact on the user throughput at RLC level.

$$\text{Average RLC user throughput } (\Gamma) = \kappa \cdot N_u \cdot RF \tag{1}$$

where
- κ is the amount of data traffic that can be transmitted with a fully utilized TSL (1 Data Erlang), i.e. timeslot capacity.
- N_u is the number of timeslots that can be allocated to a mobile, i.e. multislot terminal capability.
- RF is a reduction factor that varies between 0 and 1, which takes into account the effect of TSL sharing among (E)GPRS users. The RF in a very saturated network will be close to 0, whereas in a low loaded network it will be close to 1. Therefore, it can be considered as a measure of the (E)GPRS background service congestion in the same way that blocking probability is used for speech services.

TSL capacity (κ) and RF concepts are further described in the following sections.

TSL Capacity

Timeslot capacity is a measure of how efficiently hardware resources are being utilized to transmit data. It is defined as the data per time unit (kbps) that can be carried over one data channel (PDCH). An in-depth description can be found in section 5.2.2.7 of [3].

Timeslot capacity mainly depends on network interference levels, but it is also affected by some RLC protocol effects (e.g. the use of pre-emptive transmissions, polling strategy, RLC window stalling, etc.). The following figures show the timeslot capacity for EGPRS, assuming pre-emptive retransmissions and Typical Urban scenario (TU3) for different frequency reuses. As shown in Figures 5.27 and 5.28, when the network load increases the timeslot capacity decreases because of higher interference levels. In the case of BCCH layer (Figure 5.29), the interference is constant since BCCH TRX is always transmitting at full power, and therefore the timeslot capacity is not affected by the load.

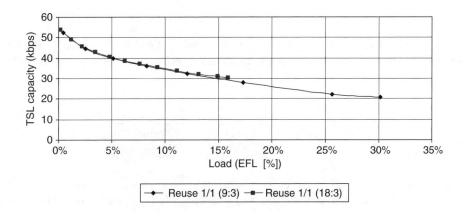

Figure 5.27   EGPRS TSL capacity for hopping layer

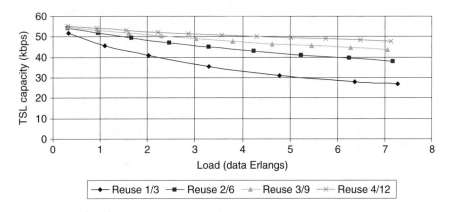

Figure 5.28   EGPRS TSL capacity for non-hopping layer

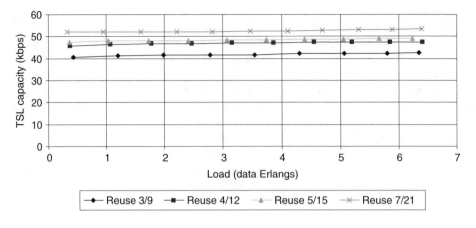

Figure 5.29   EGPRS TSL capacity for BCCH layer

In Figure 5.27, *x*-axis represents the EFL, which is a measure of how loaded a frequency of a sector is, i.e. how much traffic is carried by the available spectrum (see section 5.3 of [3]). This measure can be computed as follows:

$$\text{EFL }(\%) = \frac{\text{Erlangs}_{\text{BH}}}{\text{Total\_nbr\_frequencies}} \times \frac{1}{\text{Ave}\left(\dfrac{\text{nbr\_TSLs}}{\text{nbr\_TRXs}}\right)}$$

The estimation of the TSL capacity is one of the first steps for network dimensioning since it tells how much data the network can transmit with a fully utilized TSL. This value is needed to calculate the number of resources (TSLs) that are needed to support a determined data load. However, dimensioning for PS services not only considers the PS load that needs to be supported but also other requirements, if a minimum quality of service shall be provided.

As it was commented before, TSL capacity depends mainly on the interference levels, but not on hardware dimensioning. Therefore, it cannot be improved by increasing the resources per cell (number of TRXs) but by other network planning aspects like frequency planning or site location. However, the estimation of the TSL capacity is essential to understand the end-user performance.

Reduction Factor

Dimensioning hardware resources for achieving a certain cell data capacity (kbit/sec/cell) is straightforward, as throughput per TSL tells how much the network can transmit on each timeslot. However, this kind of dimensioning does not say anything about the user average throughput. If a minimum user quality criterion is to be considered, it is necessary to dimension the hardware resources in order to achieve this criterion. In order to do this planning, it is required to be able to estimate the RF to achieve a certain user throughput, given a certain PS and CS load.

As it was exposed in Equation 1, user throughput depends on MS multislot capability (number of TSLs that can be allocated to the mobile), throughput per TSL (that depends on

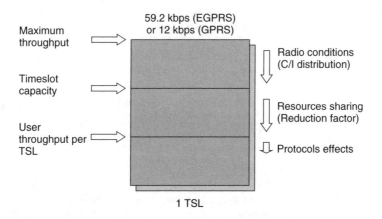

Figure 5.30 User throughput per TSL

the radio link conditions), and users multiplexing, measured in the RF figure. In addition to this, user throughput is also degraded by higher protocol effects, e.g. TCP slow start and retransmissions (as explained in section 5.2). The following Figure 5.30 shows the reduction in throughput per TSL due to these factors.

Reduction Factor modelling is explained in detail in [3] (Appendix B), where no admission control mechanism in PS was assumed, so all the incoming PS calls sharing the available resources are accepted. The following figures assume that there is admission control mechanism and only 9 TBFs per available TSLs are permitted.

This model considers radio link conditions, MS capabilities, configuration of resources, traffic mix and traffic load (including half-rate traffic) in a cell level. With this model, it is possible to compute average RF as well as RF distributions, which are needed to compute, for example, minimum user throughput. Figure 5.31 shows the accuracy of the model for average RF, comparing values from the model with measurements from a real network.

However, there are many factors that affect the RF. The following curves in Figure 5.32 show how RF is affected by some of the main ones, i.e. the voice traffic, the half-rate penetration,

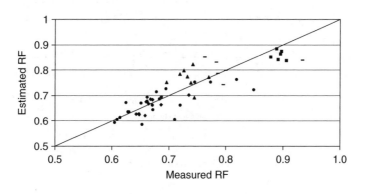

Figure 5.31 Validation of RF using real measurements

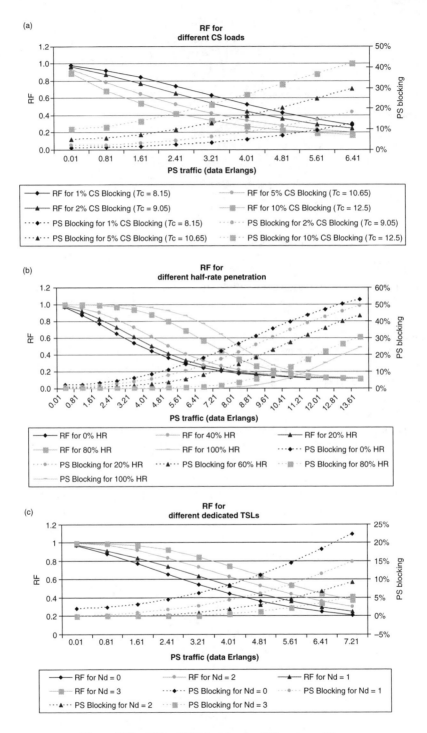

Figure 5.32 RF and PS blocking for different conditions

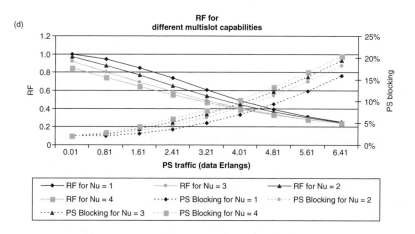

Figure 5.32 (*continued*)

the multislot capability and the number of TSLs that can be used only for PS services. These figures assume that the cell contains two TRXs.

Figure 5.32a shows the results for a typical 2 TRXs configuration with 15 TCHs. It exposes the RF (in solid lines) and PS blocking (dash lines) curves for different CS load (*Tc*). The *x*-axis represents the PS load (data Erlangs). It can be observed that RF decreases with higher CS traffic (*Tc*), since less resources are available for PS. In a very low PS load conditions (left part of the graph) the reduction is close to 1, i.e. very low multiplexing, whereas its value is decreased as PS load increases.

As it was commented, the RF is degraded for higher voice traffic, since fewer resources are left for PS services. However, Figure 5.32b reveals that RF improves when part of the traffic is carried over half-rate TCHs, since more resources are available for PS. The higher half-rate penetration, the lower multiplexing effects. Figure 5.32c demonstrates that RF is affected by the number of TSLs in the system exclusively dedicated to PS traffic (Nd). As Nd increases, the RF improves as well as the blocking figure decreases. Finally, Figure 5.32d presents how the multiplexing effects decreases, i.e. RF is higher, using mobiles with higher multislot capability (Nu), i.e. more TSLs to receive data.

The Reduction Factor figure presented in the previous section provides a characterization of the throughput degradation due to resource sharing. Thus, an estimation of the RF probability function can be mapped onto an end-user RLC throughput probability function.

End-user RLC throughput is a proper indicator of the performance for background-type of services (FTP, Web browsing, etc.). However, for services requiring a minimum user throughput (real-time services) the RF characterization is not sufficient as performance indicator.

For example, let us assume an audio-streaming service requiring 20 kbps as minimum RLC user throughput. Dimensioning of radio resources for an average reduction factor, so that average RLC user throughput is higher than 20 kbps, will not guarantee the success of the streaming sessions due to the burstiness of that RLC user throughput; for example, the user may have 30 kbps half of the time and 10 kbps the other half of the time.

The *Grade of Service* figure is introduced to take into consideration this throughput burstiness for a proper dimensioning of resources. Grade of service is defined as the probability of having an RLC throughput lower than the minimum required for the duration of each real-time session. Thus, dimensioning for a grade of service lower than 2% will guarantee that 2% of the sessions will receive at a certain time an instantaneous RLC throughput lower than the minimum required one.

Example of Radio Interface Dimensioning

The following is a theoretical example to expound the previous concepts and the methodology to be followed. Let us assume a cell with 2 TRXs in a non-hopping layer, with a frequency reuse of 1/3 and 4 Erlangs PS, i.e. 4 TCHs used for PS services in average. From Figure 5.28, the throughput per TSL for an EGPRS mobile in that cell would be around 33 kbps.

However, the effect of users multiplexing has to be considered. Assuming that CS traffic (Erlangs) is the one that causes 2% of blocking using 15 TCHs, and that 40% of the traffic load is carried over half-rate TCHs, Figure 5.32 can be utilized to estimate the RF value. With the previous 4 Erlangs PS, the RF would be around 0.65.

In order to estimate the user throughput (at RLC level), an EGPRS mobile with 3 TSLs is assumed (multislot capability). Then, user throughput could be calculated:

$$\text{Average RLC user throughput} = 33 \cdot 3 \cdot 0.65 \approx 64 \text{ kbps} \qquad (2)$$

5.5.1.2 RAN Dimensioning

In this subsection, the main RAN dimensioning issues affecting end-user service performance are analysed. They can be summarized as:

- A-bis interface, considering both control plane (TRX signalling) and dynamic A-bis issues
- BSC limitations, including specific Packet Control Unit (PCU) limitations.

A-bis Interface

Even defined as open interface by GSM standard [20], A-bis interface[11] is normally very much influenced by each specific vendor configuration, introducing vendor-dependent features. The main aspects of the A-bis interface having impact on service performance are:

A-bis Signalling: Each TRX will have a control channel associated with the A-bis interface, for proper signalling transmission. Depending on each vendor configuration, signalling resources can be shared between TRXes, or a fixed capacity LAP-D link (16 kbps, 32 kbps, 64 kbps) can be setup for each specific TRX signalling. Normally, the TRX supporting the BCCH, the CCCH and the SDCCH channels is the one requesting more signalling capacity. From service performance point of view, an improper dimensioning of the A-bis signalling resources per TRX may derive from potential problems to set up new connections and sessions (voice and data service blocking). An specific critical issue on this area is the transmission of short messages (SMSes) at peak traffic demands (e.g. on New Year's Eve): even if proper signalling capacity is guaranteed on the SMS server and radio interface, it may be the case that TRX signalling is the bottleneck.

---

[11] A-bis interface is based on PCM technology: the PCM channels (64 kbps) are divided into 16 kbps PCM subchannels.

A-bis Support for GPRS CS3–4 and EGPRS: With respect to the TCHs, the A-bis dimensioning for GSM services and GPRS CS1–2 is quite straightforward: one radio channel is directly mapped onto an A-bis PCM subchannel (16 kbps). This direct mapping is possible due to the required throughput per radio TSL for voice connections (13 kbps for full-rate voice codecs) and GPRS CS1–2 packet connections (up to 12 kbps). However, the introduction of GPRS CS3–4 and EGPRS in the radio interface brings up a new problem from A-bis resource perspectives: EGPRS provides the possibility of using 8-PSK modulation and Incremental Redundancy (IR) retransmission mechanism; and up to nine different MCSs are defined. With this enhancement, one EGPRS radio TSL can support a bit rate of up to 59.2 kbps per radio TSL; and thus there is no possibility anymore of keeping the mapping of one radio TSL to one A-bis PCM subchannel.
Again, vendors provide different solutions to this technical enhancement.

● The first alternative is to guarantee the A-bis resources for any potential number of radio TSLs with active EGPRS connections: in this case, the mapping would be one radio TSL requiring 4 PCM subchannels (64 kbps). This alternative ensures no service performance degradation due to lack of A-bis resources, but is not reasonable from capital expenditures point of view, as it requires a 300% increase in the transport resources for A-bis interface all over the network in cases where full EGPRS support is required in all the radio channels.

● A second approach would be to configure just a part of the total number of radio TSLs as EGPRS-capable TSLs. In this case, the upgrade from 1 to 4 PCM subchannels is required only for those radio TSLs configured as EGPRS. The service performance for EGPRS users is then guarantee by defining the proper number of EGPRS radio TSLs: this can be done by using some of the performance indicators present in the radio interface section (RF, grade of service) and introducing in the dimensioning methodology the effect of having just a certain number of radio resources which are EGPRS capable.

● The third alternative is to introduce the concept of A-bis pool. Instead of allocating and reserving fixed transmission capacity according to the highest possible data rate for each radio TSL and the total available EGPRS TSLs, it is much more efficient and economically practical to define a pool of PCM channels as A-bis pool.

From service performance perspective, the correct dimensioning of the A-bis pool size becomes critical in order to avoid potential degradation of the voice and data services, such as throughput degradation. On this A-bis pool dimensioning process, it has to be taken into account not only the EGPRS PS traffic (i.e. the one utilizing the A-bis pool resources) but also the CS traffic level, as far as CS calls have priority pre-emption over the PS connections.

Reference [21] includes a detailed process on how to make a proper dimensioning of the A-bis pool size. On the A-bis pool utilization, two main effects have to be taken into consideration.

1. PS radio territory usage – basically depends on the EGPRS (PS) traffic in the cell, and also on the CS traffic in the cell, due to the priority pre-emption of the CS over the PS connections.

2. Distribution of the MCS usage – the higher the MCS used by the radio timeslot, the more resources (PCM subtimeslots) taken from the A-bis pool.

BSC and PCU

On the end-to-end analysis, the BSC and specifically the PCU may introduce some specific issues on service performance.

- BSC may introduce some potential degradation on voice service, such as call blocking, if a proper dimensioning is not done. Some issues to be considered on the BSC dimensioning is the maximum number of Erlangs supported in the busy hour, and the maximum simultaneous call attempts.
- PCU limitations may introduce further throughput degradation or even PS sessions blocking. Depending on the selected PCU implementation, different issues may arise.
  - Potential lack of PCU processing capacity for the offered PS traffic
  - Potential lack of PCU channels, in case of implementing any kind of pooling mechanism with the PCU resources
  - Incorrect configuration of the Frame Relay Bearer Channels (FRBC): each FRBC has associated a certain maximum Committed Info Rate (CIR). In case the configured CIR is below the real capacity demanded from PCU/SGSN, then there is a risk of potential throughput degradation or even packet loss.

5.5.1.3 CN Dimensioning

The CN may also introduce service degradation if proper dimensioning is not done.

On the CS side, the MSC capacity needs to be estimated in terms of Erlangs supported and number of subscribers IMSI attached, so that service unavailability is avoided.

On the PS size, the capacity of the SGSN and GGSN in terms of maximum number of supported PDP contexts, processing capacity of GPRS attached subscribers have to be calculated to prevent potential PS service degradation.

5.5.2 Dimensioning Methodology

In previous sections, the potential dimensioning problems which may affect on the end-user service performance have been presented. This section summarizes those effects and describes a suitable dimensioning methodology to be followed, so that proper dimensioning of network resources is done to avoid service degradation due to network bottlenecks. Even if the following methodology is still focused on the (E)GPRS network dimensioning, the general guidelines are also applicable to any network technology.

In general, the dimensioning methodology should always be 'from bottom to top'; i.e. starting with a cell-by-cell dimensioning, then A-bis dimensioning, PCU and BSC dimensioning . . . up to the CN dimensioning. The whole process can be summarized by the following steps (Figure 5.33).

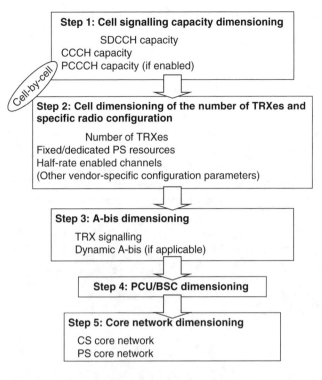

Figure 5.33   Dimensioning methodology (GPRS example)

Step 1   *Cell Dimensioning of the Signalling Capacity*

The estimation of the required resources for the signalling channels on the radio interface includes:

- Estimation of the number of radio blocks for Common Control Channel (CCCH), checking the possibility of using the combined or non-combined configuration
- Estimation of the required number of radio timeslots for the Slow Dedicated Common Control Channel (SDCCH).

A first rough estimation of these figures can be done by using traditional Erlang-B formula, considering the procedures generating load of each of the signalling channels (for SDCCH: call setup, SMSes and location updates; for CCCH: paging, procedures requesting an SDCCH, and even TBF establishment in idle mode when no PCCCH is available). Other advanced mechanisms for the signalling channels dimensioning are presented in [3].

The target of the signalling channels dimensioning is to avoid signalling channel blocking. Normally, a good dimensioning approach is to dimension a 0.2% blocking, 10 times lower than the typical blocking target for the voice channels, so that the blocking on the signalling channels is irrelevant in comparison with the blocking in voice channels.

Step 2 *Cell Dimensioning of the Number of TRXes and Specific Radio Configuration*

Once the signalling radio channels are configured, the next step is to estimate the number of TRXes per cell. This mainly depends on the offered traffic per cell (voice, data) and the grade of service requirements (voice blocking probability, average/percentile (E)GPRS RLC user throughput, Streaming Grade of Service, etc.).

Normally, the TRX dimensioning is done in two steps.

1. Dimensioning based just on load premises – where the number of TRXes to support the CS load (with certain blocking probability) and process the offered PS load is estimated.
2. Dimensioning based on service requirements – where the estimation of the number of TRXes is refined by applying the different quality criteria (RLC user throughput, streaming, grade of service, etc.). RF estimation figure with mechanisms as the one presented in [21].

In addition to the number of TRXes, the relevant configuration has to be done. Depending on the network vendor, specific network configuration parameters and features have to be configured, like for example:

- Number of fixed radio channels (i.e. fully dedicated to PS connections)
- Number of half rate capable radio channels
- Number of (E)GPRS enabled radio channels.

Step 3 *A-bis Dimensioning*

Once the cell characteristics and capabilities are properly dimensioned, the next step is the dimensioning of the A-bis interface configuration and capacity.

- For the TRX signalling, estimation of the total signalling load per TRX can be performed. With this estimation, proper dimensioning of the signalling links capacity can be performed.
- In case of using A-bis pool, dimensioning of the A-bis pool size for a required maximum A-bis pool congestion probability should be performed. Methodology as presented in [20] can be used for this purpose.

By dimensioning the TRX signalling capacity and the A-bis pool sizes, proper estimation of the number of PCM links (E1/T1s) can be done.

Step 4 *PCU and BSC*

By aggregating the traffic and signalling contributions from different cells connected to BSC, the next step is to proper dimensioning of the number of PCUs per BSC and of the BSC capabilities.

Step 5 *Core Network*

Finally, the CN has to be dimensioned to avoid potential service degradation.

- On the CS side, dimensioning of the MSC and especially of the transcoding units has to be performed.
- On the CS side, dimensioning of the SGSN and GGSN.

In general, dimensioning of the CN elements depends not only on the offered traffic but also on the connectivity to external networks, etc.

References

[1] R. Ludwig and R. H. Katz, 'The Eiffel Algorithm: Making TCP Robust Against Spurious Retransmissions,' *ACM Computer Communication Review*, Vol. 30, No. 1, January 2000.

[2] 3GPP TS 44.060, 'Radio Link Control/Medium Access Control (RLC/MAC) protocol', Release 4, V5.0.0, February 2002.

[3] T. Halonen, J. Romero and J. Melero, 'GSM, GPRS and EDGE performance', 2nd edition, England: Wiley, 2003.

[4] H. Holma and A. Toskala, 'WCDMA for UMTS', Wiley, 2000.

[5] K. Helmersson *et al.*, 'Performance of Downlink Shared Channels in WCDMA Radio Networks', Vehicular Technology Conference (VTC), Spring 2001, Vol. 4, pp. 2690–2694.

[6] Network Working Group, 'RFC 2988. Computing TCP's Retransmission Timer', November 2000.

[7] J. Torreblanca, G. Gómez and R. Cuny, 'Impact of RTT on TCP Performance over (E)GPRS Mobile Networks', IASTED CSN'2004, Marbella (Spain).

[8] R. Ludwig, A. Gurtov and F. Khafizov, 'TCP over Second (2.5G) and Third (3G) Generation Wireless Networks', RFC 3481, February 2003.

[9] K. Fall *et al.*, 'Simulation-based Comparisons of Tahoe, Reno, and SACK TCP', *Computer Communication Review*, July 1996, ftp://ftp.ee.lbl.gov/papers/sacks.ps.Z.

[10] S. Lakshman *et al.*, 'The Performance of TCP/IP for Networks with High Bandwidth-Delay Products and Random Loss', Networking, IEEE/ACM Transactions on, Vol. 7, June 1997, pp. 336–350.

[11] P. Ameigeiras. 'Packet Scheduling and Quality of Service in HSDPA'. Ph.D. Thesis Dissertation, October 2003, Department of Communication Technology, Aalborg University.

[12] 3GPP TS 23.207, 'End-to-End Quality of Service (QoS) concept and architecture', Release 5, V5.9.0, March 2004.

[13] 3GPP TS 24.228, 'Signalling Flows for the IP Multimedia Call Control Based on SIP and SDP', V5.9.0, June 2004, Stage 3 (Release 5).

[14] M. Handley and V. Jacobson, 'SDP: Session Description Protocol', IETF RFC 2327, 1998.

[15] H. Schulzrinne, A. Rao and R. Lanphier, 'Real Time Streaming Protocol (RTSP)', IETF RFC 2326, April 1998.

[16] 3GPP TR 26.234, 'Transparent End-to-End Packet Switched Streaming Services (PSS); Protocols and Codecs', Release 4, V4.3.0, March 2002.

[17] R. Price, C. Bormann, J. Christoffersson, H. Hannu, Z. Liu and J. Rosenberg, 'Signaling Compression (SigComp)', RFC 3320, January 2003.

[18] T. Berners-Lee, R. Fielding and H. Frystyk, 'Hypertext Transfer Protocol – HTTP/1.0', RFC 1945, May 1996.

[19] R. Fielding, J. Gettys, J. Mogul, H. Frystyk, L. Masinter, P. Leach and T. Berners-Lee, 'Hypertext Transfer Protocol – HTTP/1.1', RFC 2616, June 1999.

[20] G. Ramos and S. Pedraza, 'A-bis interface dimensioning for EGPRS technology', IEEE – Vehicular Technology Conference (VTC), Fall 2004, September 2004.

[21] S. Pedraza, J. Romero and J. Muñoz, '(E)GPRS Hardware Dimensioning Rules with Minimum Quality Criteria', IEEE Vehicular Technology Society, Fall Conference, 2002.

[22] G. Miller, 'The Nature of the Beast: Recent Traffic Measurements from an Internet Backbone', April 1998, http://www.caida.org/outreach/papers/1998/Inet98/Inet98.html.

6

Service Performance Verification and Benchmarking

Rafael Sánchez, Manuel Martínez, Salvador Hierrezuelo, Juan Guerrero and Juan Torreblanca

6.1 Introduction

When it comes to the verification of the performance of a network, lots of different factors are considered. There are not only technical factors, which depend on the technology in place and the measurement capabilities available; but also commercial factors, which will put the measurement in a key position as a part of contractual agreements between vendors and operators.

The vendors who made the equipment need to demonstrate that their products fulfill the expectations that their customers had and paid for, and on the other hand, the operators want to verify that everything is working the best possible way and that they are able to track the performance of their network in the most efficient and cost effective way. Usually, performance levels are key part of the contract agreements and certain values have to be fulfilled in order to really success in the network deployment and roll out (Service Level Agreements, SLAs). Of course, also in mature networks it is necessary to have in place all the monitoring capabilities that will allow keeping track of the network status and foreseeing potential needs for optimization, upgrading, dimensioning or re-designing process. The accuracy of the measurement has a key role in telecommunication business, and can be in the position to decide between business success and failure.

Of course, the complexity of wireless data networks is far bigger that it was only for voice service, as the amount of services that can be carried is very large and they present very heterogeneous profiles. In data networks, the performance is affected not only by the cellular infrastructure, but also by the external networks to which they are connected, and that provide the services itself, i.e. the Internet. Traditionally, the measurement strategies used in the cellular environments included the monitoring of certain network counters, which keep track of the performance of different network elements, and drive tests that investigated the radio and service performance from the user point of view. But now, also the impact of external networks and service providers shall be taken into account (see Chapter 7 for more details).

Generally it is possible to define three basic goals for the monitoring of any network according to the OSI Management model [1].

1. Performance monitoring
2. Network fault monitoring
3. Account monitoring.

Performance monitoring: It deals with measuring the performance of the network. It tries to capture and quantify network events as well as store the information in a common database inside the Network Management System (NMS). It is very important to choose what to measure, as there are lots of different measurable things in the network. But the list should be reduced to few meaningful and cost effective elements, which typically receive the name of *network indicators*. Normally the way to look at network indicators requires long enough periods of observation, in order to understand the tendency of the measurements and establish behavioral models. The information obtained from the performance monitoring and the tendencies are usually used as reference to plan network evolution and dimensioning, but also to detect problems at different levels (i.e. SGSN, BSS, BTS, etc.). There are also efforts to define network indicators which can provide good approaches for user performance; however, network counters tend to mix information from users with very different profiles, and also aggregate results from different areas, what really limits this possibility.

Network fault monitoring: It deals with measuring the problems in the network. The fault monitoring looks at various layers of the network, so when a problem occurs, it is also necessary to differentiate which layers of the network are affected. Additionally, in order to effectively detect problems, it is required to understand what the normal behavior of the network is during an extended period of time. It is also important to understand that there are always punctual errors in the network, i.e. due to the variability of the radio conditions and network load, but it does not mean that the network is having persistent problems. It is then necessary to look not only at specific values, but also at the duration of abnormal states, and having a clear reference of what the normal expected behavior would be.

Account monitoring: It deals with how the users interact with the usage of the network. For this purpose, the network keeps track of the utilization level of the different devices of the network. This information is used not only for billing purposes, but also for predicting future network usage as an input to dimensioning calculations.

Additionally, as part of the OSI Management model, network control functionalities are also considered.

- *Configuration*: Operator needs to keep track of the necessity for installation of new hardware and software, tracking the changes in the configuration parameters and clearly understand why, how and when to upgrade the network topology. Due to the typical large scale of the networks, it is often necessary to perform configuration audits, which verify that the right parameter configurations are in place.

- *Security*: There is the need to generate, distribute and store encryption keys for services, perform backups and ensure data security.

In addition, when the scope of the wireless network is extended to data, the problem of the user-performance monitoring starts to be more similar to the way performance measurements need to be done in the Internet than to traditional wireless network monitoring.

In the Internet world, there are no standardized monitoring tools and different tools are used depending on the purpose of the measurement. Generally, most used tools are free software applications that can be deployed in servers, routers or client terminals, which allow to perform measurements at low cost and with a large scalability. Some of these applications are very simple in concept, but allow to effectively measure key user-related performance indicators.

- *Ping*: It sends Internet Control Message Packets (ICMP) to a specific node in the network, and the packets are returned to the sender. This procedure allows to measure the response time and the percentage of packet losses.
- *FTP*: This is the simplest application for transferring files from one host to another, and is commonly used for measurement of the transfer rates. It is important to note that relatively big-sized files should be used in order to obtain an accurate estimation of the bit rate. The reason is that in every FTP session there are a number of constants delays due to the establishment time of the connection and start up delays of the TCP (slow start), which should be minimized in comparison to the real transfer time.
- *Traceroute*: This application is similar to ping, and also uses the transmission of ICMP packets to measure the delays between all the hops involved in the transmission from the client host to the service provider.

There are many other applications available, both free and commercial, which monitor the transmission rates in the client hosts, or capture IP and upper layer information to allow further analysis of packet streams.

However, due to the special characteristics of wireless networks, where the end-users have very wide mobility profiles, automation of this type of measurements is quite complicated. Also the limitation of the terminal capabilities are a restrictive factor in order to implement relatively sophisticated performance measurement applications.

Considering these limitations, the best way to measure the user performance is to directly monitor one connection and measure the Key Performance Indicators (KPIs) associated to each service, as defined in Chapter 3. This is, however, a costly solution that requires going to the field and performing active measurements, which consumes lot of time and human resources. Scalability problems clearly limit the action of the active measurements, but still there are certain situations where they are applicable: first is the roll out and troubleshooting of certain areas of the network, and second, the monitoring of important areas such as facilities of important corporate customers.

The rest of this chapter will be focused on the definition and discussion of important KPIs for data networks and understanding the complexity of active measurements and planning of trial activities. A practical view of trial activities including basic considerations, planning and tools will be shown. Different examples of active measurement over different services will be analyzed. The examples will include a basic benchmarking between different technologies and services, performed over live networks in Europe, USA and South America.

Chapter 7 will introduce new ways to look at user performance that try to complement the limitations shown for network KPIs and active measurements.

6.2 Key Performance Indicators

In the telecom world, performance measurement is usually defined in terms of accessibility, retainability and quality (Table 6.1). These terms work well for circuit-switched networks, where it is easy to identify network-measurable events and ultimately counters to each of these categories. Accessibility can be defined with the blocked calls, retainability with the dropped calls, and the quality with speech frame error rate. All these measured can be taken at the radio level (i.e. Base Station Subsystem, BSS) and can be easily translated into service quality.

However, for data services, the correlation between network metrics and user performance is not as direct as in the circuit-switched case. It is much more difficult to define counters in the BSS or other levels of the network that are able to capture the perception of service quality. There are two main reasons: first, data systems have many layers of protocols; i.e. a session may have accessibility problems, but at higher levels they may be seen just as increasing access delays. Secondly, radio data bearers are typically shared among different applications, thus mixing the information from very different sources; i.e. there exist services like Web browsing, which might be really degraded by delays, while some delays on background e-mail downloading would not be so much important.

Nevertheless, data KPIs are figures that provide significant understanding about how a certain mobile communications network is performing, specially for statistical analysis and long-term evolution, and thus it is important to define proper KPIs which try to catch as much information as possible from the network [15].

There are different ways to classify the performance indicators, based on the way that they are taken, at what level they are taken, or what is the scope that they will have. From the way the performance indicators are taken, the following classification can be done.

Table 6.1 Service satisfaction criteria

KPIs	Definition
Accessibility	Covers the user capability to the access to the radio resources, i.e. call set-up or data channel assignment.
Retainability	Covers the ability to keep up a call or the data channels in a packet-switched system.
Quality	It is a measurement of how good the connection is, or how the data service is performing. The way to measure the quality will depend on the type of service and the availability measures in the network. However, there is a clear limitation to provide user perspective of the quality from network measurements. Depending on the services, different performance indicators will be available.

- *Passive KPIs* – are computed directly by the NMS of the corresponding mobile communications network, without any active action from the operator. Usually, passive KPIs are computed in hourly periods.
- *Active KPIs* – are measured on the field by a human operator with different kinds of monitoring tools that are available in the market: Driving test tools, application monitoring tools, protocol analyzer, interface analyzers, etc. These tools allow a high level of detail in the measurement, but are limited in statistical information. Several repetitions of the tests are normally needed to consider the representative measures (see also section 6.3.3).

KPIs can be further classified depending on whether they focus on network performance or on service performance.

- *Network KPIs*: Such KPIs provide understanding about how the mobile communications network is performing from the system point of view, i.e. how the radio resources are assigned, how network interfaces are utilized, how fast mobility management signaling procedures are performed, etc. Usually, these KPIs are used to monitor the network in order to identify possible bottlenecks and dimensioning issues. Most network KPIs are passive, although there are some that are active.

- *Service-based KPIs*: Such KPIs provide understanding about how each specific service is performing from the end-user point of view. In general, network KPIs have direct impact on end-user performance, however they do not provide figures that address the performance that the end-user is experimenting. For instance, the cell re-selection outage time, which is a very common network KPI, provides the number of seconds on average that the terminal is not able to receive or transmit when a cell change takes place. This network KPI is very useful in order to know how cell re-selection operation performs in a network, however it is not giving any insight about how this gap, in transmission or reception, impacts on upper level protocols such as TCP nor how this may degrade end-user performance in terms of throughput. Each service-based KPI addresses those aspects of each specific service that have relevant impact on the end-user experience, i.e. throughput, end-to-end delay, etc. Usually, these KPIs are used for carrying out benchmarking between different mobile communications networks or systems. Also, they are used in some cases to evaluate the readiness of a certain mobile communications network or system to support a specific service. Service-based KPIs are usually active.

In the remaining of this section, a review of some of the most common network and service-based KPIs that are typically used in today's live mobile communications networks will be presented.

6.2.1 Network KPIs

Network KPIs provide understanding about how a mobile communication network is performing from the system point of view. Various groups of KPIs can be defined which allow to do the following classification according to different areas of interest.

1. *Radio Access Network KPIs*: Such KPIs provide understanding about how the radio access network is performing. Radio accessibility and retainability are two typical measures which indicate how well a user can access and keep the radio resource. However, radio performance indicators shall also be considered.

2. *Core Network KPIs*: Such KPIs provide understanding about how the core network is performing. They are normally focused on measuring the load of transport interfaces and network elements, ensuring that the core network is able to support the demand of resources from the radio access network. They also provide information that helps to detect bottlenecks in the network and take decisions about dimensioning upgrades.

3. *Signaling KPIs*: Such KPIs provide understanding about mobility management procedures performance. Signaling load from different areas can be used to analyze load distribution in the network and optimize network topology in a way that the impact of mobility procedures is minimized. For instance, high number of handovers between cells belonging to different routing areas may impact on the user performance and signaling load, and might be improved by moving some cells to a different area.

The following are the different kind of aggregations that can be typically considered for each KPI.

- *Spatial aggregation* – differentiates between different levels of the topology of the network. From cell level, which provides information about specific element, to region or whole network level, as described in Table 6.2. Cell level is normally suitable for performance verification and troubleshooting, while upper levels are preferred for studying the evolution of the network in a more generic way.
- *Temporal aggregation* – provides different level of detail when looking into the behavior of certain network element. From hourly level, which provides good insights of how the network element behaves in specific load conditions, to weekly or monthly level, which provides a view of the evolution over the time (Table 6.3).

Table 6.2 Spatial aggregations

Spatial aggregation	Description
Cell level	Cell level statistics allow to focus on the performance of a specific cell and are usually used for performance verification and troubleshooting. It is usually combined with busy-hour aggregation.
BSC/RNC	Allows to look at different areas of the network from a higher level of observation. It does not allow to distinguish details, but to follow statistical tendency for certain KPIs. It is typically combined with daily or weekly aggregation.
Region/market	Provides a general overview of the region evolution. Only a summary of the main KPIs are observed, i.e. accessibility, retainability, throughput, etc.
Whole network	This level of aggregation is mainly used for management tracking and to provide a general evolution of the network. It is usually combined with monthly or weekly level.

Table 6.3   Temporal aggregations

Temporal aggregation	Description
Hourly	Shows statistics for each hour of the day. They allow to see evolution of statistics with different load levels during the day.
Busy hour	Busy hour tries to take one hour of the day when the utilization is maximum. In this way, it is intended to ensure that low-activity periods, when the meaningfulness of the measurements might not be very reliable, do not provide false impressions about cell performance. Different busy hours may be defined, e.g. *voice busy hour* or *data busy hour*.
Daily	Daily statistics try to show a short-term evolution of the KPIs. They are used to keep track of the impact of network changes and load variations, but usually they are used to make comparison at weekly basis. As these statistics normally depend on the load, often the day of the week is also important (i.e. more load during business days).
Daily busy hour	One variant of the daily statistics, which tries to show day-by-day evolution of the network, but eliminating the effect of hours of the day with low load, where the KPI values are less significant.
Weekly	This is a medium-term aggregation which can take either daily or daily busy-hour values. However, it is usually daily evolution over the week that is preferred, in order to avoid mixing different day profiles.
Monthly	The monthly statistics are normally the highest-term aggregation and tries to give a general view of the performance of the network. This kind of aggregation is typically used for high-level reporting and general follow up of the network (i.e. load evolution, accessibility, etc), as many KPIs lose their meaning. For technical analysis of the performance, lower-term KPIs are needed.

It is interesting to note that low-level aggregations will provide more resolution in the information that can be obtained from the network: load, failures, performance, etc. of specific areas. Higher-level aggregations will provide a wider view of the performance, but will lose capability to differentiate which factors are affecting more to the indicators (Figure 6.1).

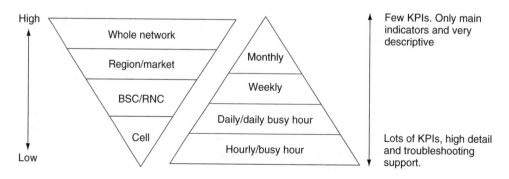

Figure 6.1   Spatial and temporal aggregations

Additionally, there are certain basic considerations to take into account when calculating or looking into KPIs at different aggregation levels. The first one is the statistical relevancy of the elements considered in the aggregation. For example, relevance of the data depends quite much on the amount of payload in the cell. Typically, if the amount of data is small, many of the KPIs may not be relevant just because there is not enough data to rely in the results of the KPI formulas (i.e. small denominators). If the aggregation over one BSC/RNC is done over a number of cells where a significant proportion does not provide relevant data, the values obtained in the higher level will also lose all its relevance. Same thing happens when making time aggregation; if daily aggregation is done over cell data which contains few hours with relevant traffic and many hours with non-relevant data, the resulting KPIs will not be representative of the performance.

In order to solve this problem, a good approach consists in discarding the information from the not reliable elements, and keeping only the information that is considered to be relevant. This is typically done from the time point of view by taking the 'busy hour' or 'daily busy hour' statistics; but from the spatial point of view the solution does not seem to be so clear. It would be possible, for example, to eliminate from aggregations all those cells which do not fulfil a minimum load level. However, doing this could have the inconvenience that some cells which are having problems or not working properly are removed from the analysis, which would not be fair, as one of the purposes of network data analysis is to provide indication of the sanity status network.

One possible solution which would allow to keep an eye on 'relevant' network elements but not forgetting about the rest, would be to define an additional KPI measuring the percentage of elements which are relevant to the performance analysis: 'Percentage of Significant Cells' (PSC). This additional KPI would provide an insight of what the weight of the cells that are not considered in the analysis because of insufficient traffic is. This way the operator can work with cells carrying enough traffic in order to perform optimization of the different indicators without being disturbed by fluctuations of the values due to random variability of the data.

Of course, cells inside the discarded group are yet first candidates for troubleshooting check. If they remain unused for a long period of time, a closer analysis would be needed to identify if there is really no data usage in that area, or if the cell is faulty.

Note that the problem of low loaded cells is especially important in the first stages of data services introduction in cellular networks, when operators have to verify that the network is working properly, but neither the services nor the users are mature enough to generate consistent data load spread along the network.

The aim of this section is to define network KPIs that can be used in all the existing live mobile communications networks no matter the technology that has been deployed, i.e. GPRS, EGPRS, WCDMA, CDMA2000, etc. In Table 6.4, all network KPIs, which will be defined in the reminder of this section, are summarized and classified.

6.2.1.1 Radio Access Network KPIs

These KPIs provide understanding about how the radio access network is performing. In this section, the most commonly used radio access network KPIs in today's live mobile communications networks are described.

Table 6.4 Network KPIs

KPI name	Active	Passive	Radio network KPI	Core network KPI	Signaling KPI
Radio resources availability	—	X	X	—	—
Radio resources allocation	—	X	X	—	—
Radio resource establishment time	X	—	X	—	—
Radio resources request rejections (blocking)	—	X	X	—	—
Number of connections	—	X	X	—	—
Successful connections ratio	—	X	X	—	—
Throughput per radio channel	—	X	X	—	—
Throughput per connection	—	X	X	—	—
Coding scheme/modulation and coding scheme distribution	—	X	X	—	—
Transmission resources utilization	—	X	X	—	—
Radio controllers utilization	—	X	X	—	—
Number of cell changes	—	X	X	—	—
Cell change service outage time	X	—	X	—	—
Serving RNS relocation service outage time	X	—	X	—	—
Soft handover success rate	—	X	X	—	—
Soft handover overhead	—	X	X	—	—
Inter-system hard handover success ratio	—	X	X	—	—
Intra-system hard handover success ratio	—	X	X	—	—
G_b interface utilization	X	X	—	X	—
I_u interface utilization	X	X	—	X	—
G_n interface utilization	X	X	—	X	—
G_i interface utilization	X	X	—	X	—
G_r interface utilization	X	X	—	X	—
GSN utilization	—	X	—	X	—
Attach delay	X	—	—	—	X
Attach failure rate	X	X	—	—	X
PDP context activation delay	X	—	—	—	X
PDP context activation failure rate	X	X	—	—	X
Intra-SGSN RAU failure rate	—	X	—	—	X
Inter-SGSN RAU failure rate	—	X	—	—	X

- *Radio resources availability*: This passive KPI provides information about the number of radio resources that are available to be allocated to new incoming connections. In case of GSM/GPRS networks, it may be number of timeslots (TSLs), while in CDMA or WCDMA it may be number of codes or available power.
- *Radio resources allocation*: This passive KPI provides information about the number of radio resources that have already been allocated to new incoming connections.

- *Radio resource establishment time*: This active KPI provides information about the time that elapses since new radio resources are requested by an incoming connection until those radio resources are allocated.
- *Radio resources request rejections (blocking)*: This passive KPI provides information about the number of radio resource requests that have been rejected by the network. Usually, this KPI is shown as a percentage of the number of rejects from the total number of resource requests, but it is also possible to present it as the number of rejections per amount of data (i.e. blocks per Erlang), or even number of rejections per unit of time (i.e. blocks per hour).
- *Number of connections*: This passive KPI provides information about the number of connections that are being currently served by the radio access network. For instance, this KPI refers to the number of Temporary Block Flows (TBFs) in GPRS/EGPRS or to number of Radio Access Bearers (RABs) in WCDMA.
- *Successful connections ratio*: This passive KPI provides information about the percentage of the connections that end successfully from the total number of established connections.
- *Throughput per radio channel*: This passive KPI provides information about the throughput that is achieved per radio channel. Typically, this KPI will provide a view of channel throughput availability, without considering what would be the final performance perceived by different terminals.
- *Throughput per connection*: This passive KPI provides information about the throughput that is achieved per connection. There are different ways to estimate the user throughput from network counters, however, in order to consider different kind for users, with different Quality of Service (QoS) profiles, the network will have to implement different categories of KPIs. Otherwise, the information will be mixed and the meaningfulness will be lost.
- *Coding scheme/modulation and coding scheme distribution (GPRS/EGPRS specific)*: This passive KPI provides statistical information about how frequently each GPRS and EGPRS modulation and coding scheme is used.
- *Transmission resources utilization*: This passive KPI provides information about the usage of transmission resources. Depending on the actual technology, one transmission resources utilization KPI can be defined for each of the following interfaces (Figure 6.2).

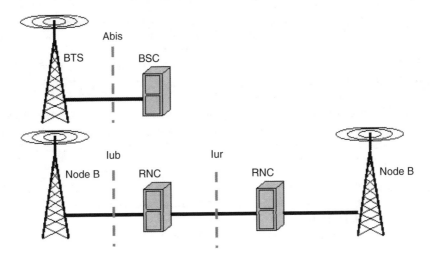

Figure 6.2   Main radio access network interfaces

- *Abis interface utilization (GPRS/EGPRS specific)*: Ratio of the used resources to the total capacity of the BTS–BSC links.
- *Iub interface utilization (WCDMA specific)*: Ratio of the used resources to the total capacity of the Node-B–RNC links.
- *Iur interface utilization (WCDMA specific)*: Ratio of the used resources to the total capacity of the RNC–RNC links.

- *Radio controllers utilization*: This passive KPI provides information about how much traffic the radio controllers are serving. Depending on the technology, the following KPIs can be defined.

  - BSC/PCU utilization (GPRS/EGPRS specific)
  - RNC utilization (WCDMA specific).

- *Number of cell changes*: This passive KPI provides information about the number of cell changes that take place. It is interesting to identify how many of the total number of cell changes take place between different routing areas.
- *Cell change service outage time at radio level*

  - *Intra-SGSN, intra-RA cell change service outage time (GPRS/EGPRS specific)*: This active KPI provides information about the duration of the data transmission gap that takes place when a cell re-selection is performed.
  - *Intra-SGSN, inter-RA cell change service outage time (GPRS/EGPRS/WCDMA specific)*: This active KPI provides information about the duration of the data transmission gap that takes place when a cell change between two different RAs within the same SGSN is performed.
  - *Inter-SGSN, inter-RA cell change service outage time (GPRS/EGPRS/WCDMA specific)*: This active KPI provides information about the duration of the data transmission gap that takes place when a cell change between two different RAs that belong to different SGSNs is performed.

- *Serving RNS relocation service outage time (WCDMA specific)*

  - *Intra-SGSN, Intra-RA serving RNS relocation service outage time*: This active KPI provides information about the duration of the data transmission gap that takes place when a serving RNS relocation procedure is performed within the same RA.
  - *Intra-SGSN, Inter-RA serving RNS relocation service outage time*: This active KPI provides information about the duration of the data transmission gap that takes place when a serving RNS relocation procedure is performed and the RA has changed.
  - *Inter-SGSN, Inter-RA serving RNS relocation service outage time*: This active KPI provides information about the duration of the data transmission gap that takes place when a serving RNS relocation procedure is performed and the new RA where the UE is located is managed by a different SGSN.

- *Soft handover success rate (WCDMA/CDMA2000 specific)*: Is a passive KPI which provides the proportion of successful soft handovers performed over the total amount of tried handovers in the CDMA or WCDMA network.
- *Soft handover overhead (WCDMA/CDMA2000 specific)*: This passive KPI provides a quantification of the soft handover activity in the network. In general, the soft handovers

increase the interference in the network which decreases network capacity. Therefore a compromise should be reached between soft handover diversity gain and interference generation.

- *Inter-system hard handover success ratio (WCDMA/CDMA2000 specific)*: This passive KPI provides the percentage of the cases in which inter-system hard handover is successful.

6.2.1.2 Core Network KPIs

Core network KPIs provide understanding about how the core network is performing. The core network is a fixed transport network, where the most important task is to monitor the load utilization of the different interfaces and network elements in order to ensure that no bottlenecks limit the potential performance of the radio network (Figure 6.3). The KPIs defined at this level must measure the capacity and utilization of the elements and must be used as reference for dimensioning activities, in combination with traffic load forecasts. In this section, the most commonly used core network KPI in today's live mobile communications networks will be described.

- G_b *interface utilization (GPRS/EGPRS specific)*: Ratio of the used resources to the total capacity of the BSC–SGSN link. The nature of this KPI is both passive and active. It is normally measured with network counters, but it can be also monitored by protocol analyzers in order to obtain more detailed information of the flows that are transported through the interface.
- I_u *interface utilization (WCDMA specific)*: Ratio of the used resources to the total capacity of the RNC–SGSN link. The nature of this KPI is both passive and active.
- G_n *interface utilization (GPRS/EGPRS/WCDMA specific)*: Ratio of the used resources to the total capacity of the SGSN–GGSN link. The nature of this KPI is both passive and active.

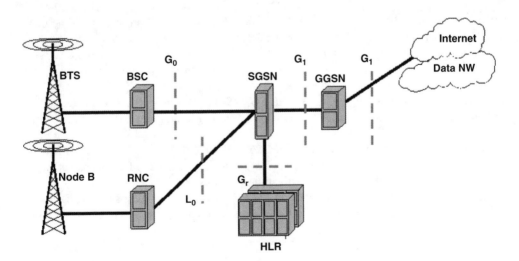

Figure 6.3   Main core network interfaces

- G_i *interface utilization (GPRS/EGPRS/WCDMA specific)*: Ratio of the used resources to the total capacity of the GGSN–PDN (Packet Data Network) link. The nature of this KPI is both passive and active.
- G_r *interface utilization (GPRS/EGPRS/WCDMA specific)*: Ratio of the used resources to the total capacity of the SGSN–HLR link. The nature of this KPI is both passive and active.
- *GSN utilization (GPRS/EGPRS/WCDMA specific)*: Both SGSN and GGSN utilization may be measured in terms of the number of subscribers that they have to handle. Several KPI are defined for this purpose.

  - *Number of attached subscribers*: This passive KPI provides the number of users attached to the GPRS/EGPRS/WCDMA network. This value is very important to evaluate the need for new SGSN elements in the network, as the number of simultaneous attached subscribers is a hard limit for the SGSN.
  - *Number of active PDP contexts*: This passive KPI provides the number of PDP contexts that have been generating traffic.
  - *Number of attach attempts*: This passive KPI provides the number of attach trials.
  - *Number of PDP context activation attempts*: This passive KPI provides the number of PDP context activation trials.
  - *Number of RAUs*: This passive KPI provides the number of routing area updates that takes place in each SGSN.

6.2.1.3 Signaling KPIs

Signaling KPIs provide understanding about how mobility management procedures perform. In this section, the most commonly used signaling KPIs in today's live mobile communications networks will be described.

- *Attach delay (GPRS/EGPRS/WCDMA specific)*: This active KPI provides the time that it takes for a mobile terminal to attach to the network.
- *Attach failure rate (GPRS/EGPRS/WCDMA specific)*: This KPI, which may be active or passive, provides the ratio of the number of failed attaches to the total number of attach attempts.
- *PDP context activation delay (GPRS/EGPRS/WCDMA specific)*: This active KPI provides the time that it takes for a mobile terminal to activate a PDP context in the network.
- *PDP context activation failure rate (GPRS/EGPRS/WCDMA specific)*: This KPI, which may be active or passive, provides the ratio of the number of failed PDP context activations to the total number of PDP context activations attempts.
- *Intra SGSN RAU failure rate (GPRS/EGPRS/WCDMA specific)*: This passive KPI provides the ratio of the number of failed RAUs to the total number of RAU attempts within the same SGSN.
- *Inter SGSN RAU failure rate (GPRS/EGPRS/WCDMA specific)*: This passive KPI provides the ratio of the number of failed RAUs to the total number of RAU attempts between two different SGSNs.

6.2.2 Service-Based KPIs

Service-based KPIs provide understanding about how each specific service is performing from the end-user point of view. Each KPI addresses those aspects of specific service that have relevant impact on the end-user experience. In Table 6.5, all service-based KPIs, which will be defined in the reminder of this section, are summarized and classified.

6.2.2.1 FTP Service

- *FTP start-up failure rate*: This active KPI provides the percentage of cases in which the FTP client is not able to connect to the server (control connection failure), or it is not possible to establish the data TCP connection for transmitting the file. Causes of failure maybe related to network conditions, TCP/IP level connectivity or application server problems, i.e. congestion. Also, in some situations firewalls can limit the connectivity to FTP service.
- *FTP abort rate*: This active KPI provides the percentage of cases in which FTP transmission is started, but is aborted before file is fully transmitted. Causes of failure maybe related to network conditions, TCP/IP level connectivity or application server problems, i.e. congestion.
- *FTP throughput*: This active KPI provides the FTP throughput that is measured when the file has been fully transmitted, and is directly related to the quality that end-user experiences. The throughput is provided at application level. This KPI provides the average throughput over the whole FTP session.

Table 6.5 Service-based KPIs

KPI Name	Active	Passive
FTP start up failure rate	X	—
FTP abort rate	X	—
FTP throughput	X	—
HTTP access failure rate	X	—
HTTP abort rate	X	—
HTTP access time	X	—
HTTP access time to text	X	—
HTTP throughput/delay	X	—
PoC service availability	X	—
PoC service accessibility	X	—
PoC voice quality	X	—
PoC timely delivery of voice	X	—
MMS send/retrieve failure rate	X	—
MMS send/receive throughput	X	—
MMS send/receive delay	X	—
MMS end-to-end delay	X	—
MMS notification delay	X	—
WAP failure rate	X	—
WAP access time	X	—
Ping (Round-Trip Time, RTT)	X	—

6.2.2.2 WWW Service (HTTP)

- *HTTP access failure rate*: This active KPI provides the percentage of cases in which the HTTP client is not able to connect to the server. Causes of failure maybe related to network conditions, TCP/IP level connectivity or application server problems, i.e. congestion.
- *HTTP abort rate*: This active KPI provides the percentage of cases in which a HTTP transmission is started but is aborted before the page is fully loaded. Causes of failure maybe related to network conditions, TCP/IP level connectivity or application server problems, i.e. congestion.
- *HTTP access time*: This active KPI provides the delay from the moment when the user clicks on the desired Web page link to the moment when the user actually sees the first contents of the page. Main content is typically text. Images, applets, JavaScripts, etc. are loaded afterward.
- *HTTP access time to text*: This active KPI provides the delay from the moment when the user clicks on the desired Web page link to the moment when the user actually sees the full text loaded in the browser, i.e. only the text, just the .html file, no objects.
- *HTTP throughput/delay*: This active KPI provides the delay from the moment when the user clicks on the desired Web page link to the moment when the user actually sees the full page loaded in the browser.

6.2.2.3 Push-to-Talk over Cellular (PoC)

- *PoC service availability*: This active KPI provides the percentage of the cases in which PoC service is available, i.e. PoC server is up and running and it is able to establish PoC sessions.
- *PoC service accessibility*: This active KPI provides the percentage of the cases in which a PoC user is able to access to the server and establish a session for voice transmission.
- *PoC voice quality*: This active KPI provides a subjective measurement that determines the user experience with the service.
- PoC timely delivery of voice

  - *Right-To-Speak (RTS) delay*: Is defined as the time between the instant a PoC subscriber initiates a PoC session and when it receives a 'right-to-speak' indication.
  - *Start-To-Speak (STS) delay*: Is defined as the time between the instant a PoC participant initiates a floor request on an ongoing session and when it receives a 'start-to-speak' indication.
  - *Voice Delay Time (VDT)*: This active KPI provides the delay from the moment when the calling party starts speaking to the moment when the called party hears the initial speech burst.
  - *Round-Trip Time (RTT) delay*: Defined as the time since a message is sent from the calling party, till an answer is received from the remote end.
  - *PoC end-to-end delay variance in a group call*: This active KPI provides the time difference between the first reception and the last reception of a speech burst within a group call.

6.2.2.4 Multimedia Messaging Service (MMS)

- *MMS send/retrieve failure rate*: This active KPI provides the percentage of the failed send/ receive transactions from all send/receive transactions.
- *MMS send/receive throughput*: This active KPI provides the send/receive data rate in kbps.
- *MMS send/receive delay*: This active KPI provides the delay from the moment when WAP connect event takes place to the moment when WAP disconnect event takes place for the sender/receiver.
- *MMS end-to-end delay*: This active KPI provides the delay from the moment when WAP connect event takes place for the sender to the moment WAP disconnect event takes place for the receiver.
- *MMS notification delay*: This active KPI provides the delay from the moment when WAP disconnect event takes place for the sender to the moment when WAP connect event takes place for the receiver.

6.2.2.5 WAP

- *WAP failure rate*: This active KPI provides the percentage of cases in which the WAP client is not able to connect to the server.
- *WAP access time*: This active KPI provides the time it takes to download an entire WAP page.

6.2.2.6 Ping

- *RTT*: This active KPI provides the time that it takes for different packet sizes to go from the client to the server and back. This KPI therefore characterizes the latency of the network.

6.3 Trial Methodology

The purpose of a trial activity is to perform active measurements on the cellular network that will allow verifying the correct behavior of the system, testing the performance of the terminals, and discovering and/or investigating possible problems in the network. How the test are defined and implemented is not really in essence very different for a wireless data network than for a voice cellular network. There is one clear common factor in both cases, which is the quality of the signal that is received and the coverage on different areas. The principal difference will mainly come from the dynamics of the services that shall be supported. Depending on the purpose of test the following classification could be done.

- *Network coverage and radio quality*: These tests can be performed by testing devices which measure the signal strength and interferences on different areas of the networks or driving routes. This information is typically used as feedback for the radio planning activities so real propagation data can be included in the design and optimization phases of the network. This kind of measurement is typically more important during first phases of network implementation.

- *Basic service performance benchmarking*: This kind of test is normally firstly done in lab and then in the first phases of the network deployment, to verify the quality of the service in place. In case of data services these tests focus on looking at specific service-related measurements, and combination of different radio qualities. In this phase, only basic services and main KPIs are used: i.e. throughput, latency, RTT. These tests shall also include monitoring of the basic signaling procedures dealing with network attachment and terminal mobility. Different tests shall be done for each service that will be supported by the network. In each case, the performance indicators to consider may be quite different.

- *Detailed service performance verification*: This test targets to verify the performance of the network in a wider number of cases, and for several services. The objective is to verify the fulfillment of the basic performance, usually included in SLAs. Depending on the volume of the measurement campaign it will be necessary to control the number of services and KPIs to be monitored in order to keep the duration of the tests under reasonable limits. Again, most representative services and KPIs will include throughput, latency and RTT.

- *Single service verification*: These tests focus on the verification of the performance for one specific service. It is done especially for new or more demanding services which require certain minimum performance in order to fulfill user expectative, i.e. new delay sensitive services, where user satisfaction is more demanding: streaming, PoC, gaming, etc.

- *Troubleshooting*: These tests are focused on detecting and investigating problems in the whole network or specific areas. Typically, some strange behavior or unexpected values are observed from network KPIs, which require further observation in order to understand their causes and how to solve them. For general purposes, basic services to measure throughput and latency will be used unless those problems affect a specific service. In order to have a complete view of the source of problems, the information reported by field testing tools in the user side (radio quality, application-level KPIs) may be completed with information monitored from network KPIs, and/or active monitoring of application servers, network elements (i.e. SGSN, base station) or network interfaces.

- *Application performance audit*: The purpose in this case is to evaluate the overall performance of different services in the network. The main objective is to understand the user perception of the services, thus no deep analysis or control of the radio configuration is needed. The idea is being able to analyze a wider area or a larger range of service, by reducing the complexity and details of the measurements. Instead of monitoring different levels of the network and obtaining all the detailed radio information, it only looks at the user-perceived performance. Of course, this limits the possibilities to understand the reasons for degraded performance. If problems are detected, troubleshooting tests should be performed with specific services and in selected areas in order to obtain any information that is needed for the analysis. One additional scope for performance audit is the benchmarking between different networks or even technologies. As the tests do not require deep knowledge of the implementation, just availability of terminals and common application level tools (i.e. mobile device working as a modem, plus application clients running on a laptop), it is possible to get fully comparable measurements.

From a general point of view, when dealing with cellular networks the first step is always to make a correct design and dimensioning of the system. It is necessary to provide enough sites to ensure a good coverage, and in case of GSM or TDMA also to provide a good frequency planning. Being able to obtain the best radio performance is the real basis to get the best service performance also from the end-user point of view.

Once the initial phase of the network deployment is accomplished and the radio performance is optimized, or at least it is inside acceptable values, the performance of data services shall be analyzed and optimized separately. As described in previous chapters, the different characteristics of each service may lead to a particular user perception, even if the radio conditions are quite good. At this stage, the active measurement perform a main role, because they are the only way to really look deep inside the dynamics of the service performance and understand which network behaviors and service parameters could be interacting to produce a performance degradation. The most typical situations to analyze are the cases whether besides a good radio performance, the user level indicators shows underperforming values due to interaction between sub-optimal service parameters and the typical high latency delays in the wireless network. In those cases, service parameter optimization will be required.

However, in other cases the limitations may come from interactions between service dynamics and the network configuration. For instance, one network configuration may have shown good performance results for basic throughput and latency measurements; but a certain service of bursty nature may suffer for constant delays due to radio bearer establishment and release. Such service would prefer a steadier assignation of radio resource than the default in order to minimize the latency and response times due to radio bearer establishment. This kind of optimization is however done at a cost, well for network resource utilization, increased interference or impact on other services, so the decision shall be taken based on which services are more sensible for the operator.

One usually difficult use case to test in a cellular network by active measurements is a 'high load test', meaning a situation where the cell capacity is very high or even exceeds the cell's expectations. This kind of test is very valuable to determine at which point the degradation suffered by the user performance is not acceptable anymore and the cell would need to be upgraded. However, the complex logistic regarding measurement devices availability and realistic traffic mix generation limits the feasibility of performing this kind of high load tests, and in most of the cases load analysis will be mostly based on network monitoring or in theoretical studies.

Finally, there is another approach for active measurements in the network, which are rather not in the scope of trialing activities, but in the area of network monitoring. It consists of inserting certain monitoring devices in key points of the network, which will be reporting samples of the system quality from the user point of view located in few priority locations (i.e. corporate customer's facilities). This approach is further described in Chapter 7.

6.3.1 Trial Phases

A trial is a complex task that requires lots of effort from the planning and resource coordination point of view. It involves lots of people performing different activities, and also from many different areas. Depending on the level of detail that is required, it may be needed to coordinate several driving test teams with people monitoring the network statistics, and field people in charge of controlling network configuration. Furthermore, it is necessary to make sure that all the equipment, servers, cars and other logistics are in place at the right time.

From a critic point of view, it can be said that active measurements (or trials) are very time-consuming, do not give the whole picture of the network behavior, and are not statistically reliable due to the limited scope that they can cover. However, they are the best tools available to analyze in detail the behavior of the network, especially during first stages of the deployment where there is not enough traffic so that statistical measurements can be used to monitor general trends. In any case, even in mature networks, they are the more direct way to measure the service performance from a user perspective.

6.3.1.1 Elaboration of the Test Plan

The first step when preparing a trial activity is to define clearly the scope and targets of the measurement campaign. This step may be summarized by answering two basic questions: What do we need to obtain from the measurements? And how are we going to do it?

The answer to the first question depends on which category (described in previous section) the trial belongs to: general audit, troubleshooting, service performance verification, etc. Once it is known, the services and metrics to be collected have to be identified.

The answer to the second question also depends on the purpose of the test. In general, if the target is to test a new feature or service, the normal procedure is first trying to test it in a lab environment, where the configuration and radio conditions are better controlled. The second step would be to perform the tests in live cells but under controlled conditions, for example, during night time at low load hours. Initially, static tests shall be done in order to minimize additional effects generated by mobility; however, once the static conditions are tested, it is also needed to include few use cases with mobility as well. Finally, tests should be performed in wider areas on normal traffic hours (even busy hour) to understand the performance in a number of 'typical' cases.

Usually the wider the scope of the tests in terms of number of services, number of cells and number of different mobility situations, less the number of metrics should be selected to monitor, not only because of the complexity of getting the measurements, but also due to the large post-processing times that would normally be required to look into all the tests in detail. However, one common practice is to get as much information as possible from the user point of view, even if it will not be initially analyzed, but keep it available for those cases in which something does not look promising, and further analysis would be required.

At the end of the day, a good plan comprising all the information required to perform the tests will be needed: services that are to be tested, metrics that will be analyzed and tools that will be used for both performing the measurements and post processing the data.

There is also one additional consideration: the plan must be realistic in terms of time frame and leave some room for repeating failed tests and to deal with unplanned issues. The main points to consider are listed below.

- Definition of the test cases
- Metrics to be analyzed
- Logistics: engineers, tools, equipment, etc.
- Methodology and configuration
- Timing of the tasks
- Expected results.

Unfortunately, the time frame is usually one of the most restrictive aspects in the planning and implementation phases. In this aspect, it is very common that time restrictions limit the possibility to repeat tests or perform a very detailed analysis. Active measurements also have a high economical cost, which usually also restricts their scope or the detail of the analysis that can be delivered.

6.3.1.2 Detailed Definition of the Test Cases

Definition of the test cases must include three main categories.

1. What to measure
2. How to measure
3. Expected results

What to measure shall include the list of services that are required to be measured, plus the metrics that are to be collected. The description may also include why they have to be measured.

How to measure shall include the test conditions: whether it will be a static case or driving environment, what will be the duration of the test, which tools will be used and how the test will be executed.

Expected Results from the test shall be included as a reference to evaluate if the test is successful or not. This reference may be based on theoretical results, simulation estimations, or even in previous tests performed in lab and other networks. These reference values will be used to determine whether the test results are satisfactory, and define whether a more detailed analysis is needed or not.

The same split can be applied to the whole test campaign, and also to each individual test. The objective is to build a task list, which may be typically written in a table, containing all the information regarding the test. Additional information about the configuration to be used and test duration shall also be included. Table 6.6 shows an example of test case definition for standard application throughput verification.

All the test cases shall be listed and ordered by priority in a list ('Test checklist') which will be used as a summary of the test and provide basic information about test status (Figure 6.4). Additional information shall also be provided about the test area: maps of the location and position of base stations, possible driving test routes and locations for static tests.

In the definition of the test shall also be included an estimation of its duration, based on the expected radio conditions, and also consider possible delays due to practical issues. A more detailed analysis of these practical issues is included in the following sections.

6.3.1.3 Pre-Trial Field Work

Before the measurement campaign starts, all the trial system components must be installed and tested in order to ensure that everything works properly. These verifications will take place during the pre-trial phase. This phase will also allow some time for getting familiar with the tools that will be in use during the trial and the testing area. During this period, in addition to verification of the tools, the exact position for the static tests and driving routes shall be determined by preliminary measurement of the radio conditions. Also, verification of

Table 6.6 Example of test case definition

What to measure	User throughput obtained by terminal X in different radio conditions.
Test duration	One hour
How to measure	

- FTP application will be used running on a laptop.
- Connect the wireless device to the laptop and use it as a modem.
- Prepare the test settings to capture the data: Bar the cell in sector under test, verify monitoring tools are working properly and network statistics are being collected.
- Start the test 5 minutes before network reporting period (usually one hour) starts.
- Start protocol analyzer tools to capture packets in different interfaces.
- Start capture of radio quality measurements.
- Perform download sessions of 1-MB file with 10 second delay between sessions.
- Repeat download 10 times.
- Perform upload session of 500 kB (i.e. if link is asymmetric), with 10 second delay between sessions. A total of 10 pings with one single packet per command.
- Finish the test 5 minutes before network reporting period ends.
- Collect data from network counters, protocol analyzer on different interfaces, and TCP/IP throughput measured.

Expected results	Under low load conditions and with a good quality the terminal X, with radio capabilities Y, is expected to provide a throughput in the radio level in the range of [thr1, thr2]. The throughput at application level, due to reductions by establishment delays, protocol delays and overheads is expected to be in the range [appThr1, appThr2].
Configuration	TCP Parameters

- Set optimum MTU and AWND size (see Chapter 5)
- Activate cell barrel
- Verify that main cell parameters fit with standard configuration that is to be tested. Verify also hardware and SW version of network and test equipment.

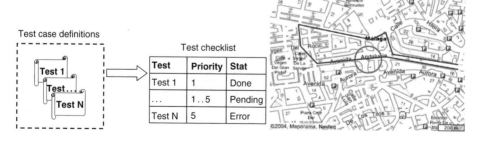

Figure 6.4 Test definition tables and driving test route

Inventory list

HW	Nw SW	Client Sw	Measurement tools
BTS model	BTS Sw Ver	Automatic test generator	Drive test tool
GGSN mod	GGSN Sw Ver	Windows/Linux X.x.x	Protocol Analyzer
Gn link capacity	FTP server	MS terminal drivers	OSS counters
Gb link capacity	Windows/Linux X.x.x	Application clients	MS terminal
—	HTTP Sever	—	—

Figure 6.5 Inventory list: HW, SW and measurement tools

the measurement tools shall be done, and in cases where different tools are available for measuring the same parameters, measurement results obtained with these tools shall be cross-checked. Those tools providing the most reliable results will be selected for use during the trial.

Typically, all the elements involved in the trial – hardware, software, testing tools, applications, etc. – should be listed in an inventory and kept for the record of the trial, as described in Figure 6.5. This would allow to further repeat the same tests (if necessary) or compare with other test environments.

6.3.1.4 Testing

The success of the testing phase will depend very much on how well the planning was done and that all the necessary actions to perform the test and capture the data are properly done.

First of all, the timing between the active measurement equipment and the measurement tools in the network must be synchronized, and the start time of the test must fit within the proper measurement window. In order to ensure that everything works properly, all the measurement tools should have been already tested during the pre-trial phase.

The number of repetitions and duration of the tests is a key factor, which should have already been planned from the test-definition phase. It is very important to have enough repetitions of each test in order to ensure some statistical relevance of the obtained results. However, the number of repetitions is normally restricted by the time constraints of the test. At the end of the day, tests taking longer time (like long file downloads) will be required to have fewer repetitions in order to fit into reasonable measurement windows. As a rule of thumb, each test should be repeated at least ten times, unless strong time constrains exist. No less that five repetitions should be done of each test.

Another important aspect, which should be also considered already at the definition of the test, is the time frame at which they would be made. Depending on the purpose of the trial it might be interesting to have measurements during night time where the load of the network is low and user can measure maximum performance offered by the network (minimize interactions with other users). But the test can also be done during busy hours, in order to understand the degradation of a user when interacting with other normal traffic in the network. These different situations may require that the same test is repeated at different times of the day, and they should be treated as independent test cases from the planning point of view. Anyhow, they shall be considered together for analysis purposes.

During the whole testing phase it is important to generate proper documentation of the test, including main information about the hardware involved, network configuration and software version, date/time, responsible person, etc. The most recommendable is to have a template table where the engineer fills the information before and after performing each test.

Finally, there is one fact that has to be always considered when planning active tests, and also when they are being performing. Due to the empirical nature of the tests, even the best-planned campaign has a high probability that something goes wrong at some point, causing extra delays. For instance, one computer may crash, an application server prevents connections to be established, terminal may have problems to connect to the network, or even a policeman may wonder what three guys are doing in a van at 4 am in the middle of a residential area. Every test plan shall consider this, and leave some room for test repetition, i.e. 10 or 20% depending on the risk that could be expected. If at the end of the day there is some spare time from the measurement activity (which is very unlikely), there will be always a lot to do to process and analyze the data.

6.3.1.5 Analysis

The analysis of the data should already start right after the tests are finalized, by checking at least few key indicators like, for example, application throughput. These few indicators should give the confidence that the test results are within reasonable values, and there is no need to repeat them. If the results are not within reasonable expected results, then the engineer should try to find out possible causes, double check the equipment and configuration, and if the time frame allows it, try to repeat some of the cases to verify that the behavior is consistent. More detailed analysis should be done afterward. This kind of sanity check should be done at least at the end of each testing day.

After this first verification phase, the results shall be presented in a progressive way. First, few main indicators shall be selected which show the overall picture of the system behavior. These KPIs must cover the performance of the radio interface, network and user perception for each of the tested services. For example, these results may present average throughput and latency measured from network indicators and mobile side, and average end-user quality for each of the services (section 6.3.2).

In order to be able to process efficiently the enormous amount of data that can be collected from extensive active testing it is important to count with post-processing tools which are able to look into the traces captured in the different tests, and automatically calculate average values, tendencies, or plot the results graphically. There are commercial tools that may help to do some of these tasks, but typically it will be needed to personalize some of the post-processing work, for example, customize measurements that are not available in the tools. In those cases, text processing tools like 'perl' or macros associated with spreadsheets are typically used.

After obtaining the general results that will draw the performance of the system, a more detailed analysis of specific cases may be included in order to better understand what happens or troubleshoot bizarre results. Only few cases may be analyzed in detail, because normally this process requires lots of effort and time; so it is important to select among the different tests which are the cases that are more significant for the analysis. Some criteria may be to take *best*, *worst* and *average* case from a whole set of tests, and compare the radio condition and upper layer analysis.

The example in Figure 6.6 shows the detailed analysis of a 512 kB FTP file download and EDGE terminal with 2 TSLs, which provided an average application throughput of

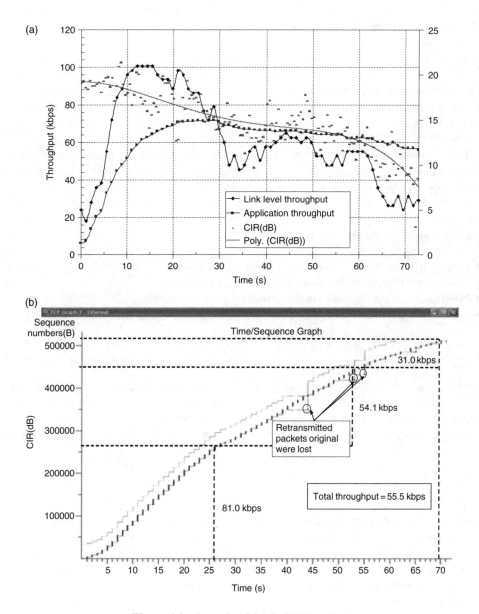

Figure 6.6   Example of detailed FTP analysis

55.2 kbps (57 kbps at radio level). It has been chosen as an intermediate case where the throughput was neither the worst possible nor the best (up to 90–100 kbps). In Figure 6.6a it can be observed the correlation between link level throughput, average application throughput and the radio quality evolution measured in the user side. It can be observed how the reduction of the radio link quality impacts directly on the degradation of the measured throughput. The same behavior can be also checked looking at the graphical presentation of the traces from the protocol analyzer in Figure 6.6b. Dividing the transmission into three shorter parts the different transmission rates can be observed at TCP level (81 kbps, 54.1 kbps and 31 kbps, Figure 6.6b), which are aligned with the radio quality evolution (Figure 6.6a). However, two additional effects that affect average throughput are observed: first, the TCP establishment delay, which is not considered in the figure but it is considered for the application throughput calculation (typically 1.5*RTT); and secondly, the TCP slow start phase, which shows an exponential growth of the throughput during the beginning of the transmission. Finally, also the presence of few TCP retransmissions, which are included in the link level throughput, but do not contribute to the application level throughput can be observed.

This analysis from the user perspective could be also completed with the monitoring of network interfaces, especially for troubleshooting. However, the complexity of this cases tends to be very high (also the time needed), and will be done very rarely.

Finally, the key aspect in the analysis must be the outcome conclusions which will include the understanding of current performance based on the measurement, and identification of possible sources of performance degradation.

- How well are the results aligned with the expected results
- Which were the reasons for under-expected values (if any)
- What cases were not covered in the analysis (i.e. high cell load)
- Proposal for improvements (i.e. software upgrades, increase resources, improve coverage, etc.).

6.3.2 Main Measurements

This section will go through some of the main performance indicators that affect the user performance in any data network, and which are especially important for wireless and cellular network due to their higher delays and variability compared to fixed lines. For each of these performance indicators, the best way to measure them by using active measurements will be also proposed.

6.3.2.1 Throughput

User throughput is, maybe, the most well-known indicator of the end-user performance. It measures how fast the user is able to obtain/send data from/to the network. Throughput is always measured as the division between data transmitted and the time needed for the transmission, thus any extra delay that is considered in the measurement (in addition to real transmission time) will reduce the throughput perceived by the end-users. For this reason, bursty kind of traffic, which tends to have several transmission gaps between transmitted elements, usually achieves low throughput when averaged over a normal session. This is the case of WAP traffic or, at lower level, HTTP traffic.

The most accurate way to measure the real throughput provided by a network from the user point of view is to use a data source providing a constant transmission rate. This is the case of FTP, which is normally used as reference for any throughput benchmarking. However, FTP also has some inherent delays, mainly due to TCP establishment and congestion control procedures (slow start) which add some constant delays at the beginning of the transmission and that will affect the average throughput. The solution is to use FTP with a large file, which will make these constant delays very small compared to the overall transmission time, thus minimizing the application throughput degradation. Optimal size to be used will depend on the bit rate available and bandwidth delay product, so systems with high bandwidth might need bigger files. Typical values used for 2.5G systems are 1 MB or 500 kB. 1 MB gives more accuracy, but may limit too much the number of repetitions (make tests very long) if the available bit rate is small.

6.3.2.2 Round-Trip Time (RTT)

RTT is the time needed for one packet to go from the terminal to a server in the Internet and back. The value of the RTT depends on the size of the packets, and is a typical indicator of the quality of the network (it affects the performance of TCP and application layers, as described in Chapter 5).

RTT is typically measured by using the 'Ping' application, in a similar way than described in previous section. Normally the RTT values are defined for a steady transmission state, which implies that the effect of the initial radio bearer assignation should not be counted. In this case, the ping delay is averaged over ten or more packets which are transmitted one after the other without any delay, in order to allow the system to keep the radio bearers assigned.

6.3.2.3 Establishment Time

At radio level, it is the time needed to establish a data connection (or radio bearer) between the terminal and the network. This time is mainly determined not only by the network architecture, but also by the existing load. It is also possible to define establishment times at different levels, such us the establishment time of a TCP connection.

In order to measure the radio bearer establishment time, the best approach is to use a *drive testing tool* (section 6.3.3), which will capture the signaling messages between the terminal and the network, and provide very accurate values. However, these tools are not always available, so some indirect approaches can be defined in order to obtain an estimation of the latency used to benchmark or compare the network performance. The most common is to use the 'Ping' command, present in many operating systems. One packet is send to an IP address, which will return to the terminal another packet of the same size. If the packet is selected to be very small (i.e. 32 bytes), the transmission time will be negligible, and the measured delay is mainly because of the time needed to establish the radio resources. Normally, it is needed to wait few seconds before repeating the test, in order to ensure that the radio bearers are released.

6.3.2.4 Cell Reselection Times

Depending on the system, there might be outage times while the terminal is changing from one cell to another. These times may change depending on the network configuration, radio conditions, coverage, etc. So, it is important to measure and understand the impact that these events may have on users. Normally, static users will not change the cell they are connected to, unless in poor or very changing interference conditions; so the most likely situations to observe cell reselection will require driving test.

Generally, in order to measure the cell reselection times between two cells, a route that goes from one cell to another is defined, following a given path on which the quality is progressively degraded. There are two typical measurements.

1. *Radio outage time*: It is the time since the radio connection in the old cell is stopped, till the radio connection in the new cell is established. This gives a measurement of the impact on the network, but cell reselections may have another radio procedures associated (like Routing Area Update) that can interact with upper layer protocols (i.e. generating spurious retransmissions in TCP [14]). In order to measure this outage, a driving test tool is needed which looks into the radio signaling information.
2. *User outage time*: It is the time since the user receives the last data packet in the old cell, till it receives the next data packet in the new cell. Note that if unnecessary TCP retransmissions are triggered due to this procedure, the real outage time from the user point of view will also include the transmission time for the retransmitted TCP segments. This outage time can be measured through traces of a TCP protocol analyzer. If the number of cases to investigate is going to be high, an appropriate post-processing tool would be required for efficient calculations.

6.3.2.5 Mean Opinion Score (MOS)

MOS measures the subjective quality perceived by a group of users. It is normally used for services whose quality is difficult to be evaluated with measurable terms such as throughput or delay. The most extended usage for MOS measurements is the speech quality, but variants are also used for network gaming experience, for example.

Normally, MOS surveys require a group of persons that experience the service under different conditions and give a grade to the quality. This kind of test is quite costly and not feasible for normal network verification or troubleshooting. The normal approach is to perform the MOS tests under controlled environments and establish a relationship between the MOS quality and one or several measurable indicators. This relationship is normally presented in form of tables or charts. This way, voice or gaming services can be later evaluated by quantifiable measurements, and mapped onto a meaningful MOS indicator. An example of this procedure can be found in section 6.4.6.

6.3.3 Tools

In order to measure the performance of the network, it is essential to have appropriated tools. These tools need to cover from the collection of network information and counters to monitor the different interfaces, and measure the quality experienced by the end-user. When performing network measurements it is important to look at all the information available in order to verify the proper behavior of the network, and in case that problems are detected, being able to troubleshoot and find the source.

However, the analysis of all the collected data is normally even more complicated and time-consuming than getting the information itself. It is really difficult for an engineering group to look in detail at all the collected data, protocol traces, etc., so in practice this is only done for troubleshooting purposes and generally only a selected subset of information is used.

In addition, other important factor to consider is that the data is not always obtained in the best understandable format for a human being. Therefore, additional post-processing tools are needed to filter out less significant information, highlight the most important, and combine automatically the data from different sources to build a consistent set of information which can be analyzed and presented to the management.

6.3.3.1 Drive Test Tools

Tools of this kind are used to capture the throughput at lower layers over the air interface, measure radio conditions, and monitor signaling messages between the terminal and the network. This kind of tool is very important for detecting specific problems in the network and performing troubleshooting. They can also be used in combination with application-level tools in order to correlate the radio performance with the end-user experience. Normally, these tools are composed of one mobile terminal with special firmware and software that is able to collect or transmit to a laptop all the information from the radio interface that is captured.

Typical information that can be obtained from these tools is as follows.

- Information about the system and serving cell: Cell Id, frequency, broadcast information, etc.
- Measurement of radio quality: received power, signal-to-interference ratio, block error rate, etc.
- Throughput and delay on the radio interface.
- Signaling messages in the radio: Attach to the network or session establishment (i.e. PDP context activation in 3GPP), etc.
- Special cellular events, like cell changes or routing area updates.
- Establishment of radio bearer or change of bearer capacity, for example, assignment of higher bit rate channel in WCDMA and CDMA or increase the number of TSLs in (E)GPRS.

Drive test tools also use to include positioning systems (e.g. Global Positioning System, GPS) in order to correlate the measurements with different locations.

The capabilities of drive test tools are very important, especially during deployment and roll out of the network, but they are also indispensable for performance verification and troubleshooting.

6.3.3.2 GPS and Digital Maps

GPS applications in combination with digital maps are very important software to be used during driving tests in order to be able to fully locate the position of the terminal during driving test routes. This information also allows to correlate the measured signal levels and other quality indicators with the position in the map, thus allowing to find out coverage and interference problems in the network and to create realistic propagation maps. Usually, GPS and digital mapping capabilities are already included in most popular drive test tools.

6.3.3.3 Automatic Application Testing and Monitoring

Automatic monitoring tools are needed to allow the engineers to define simple or complex tests by means of scripts or other kind of scheduling system. These tools will activate the wireless connection and automatically execute different services without the need of specific actions from the engineer.

This kind of tool allows to minimize the human factor in the realization of the tests, by defining all the test cases before hand, and ensuring that the tests are always done in the same order, with the same applications and with similar timing between repetitions.

These applications also include measurements of the main user KPIs related to the supported applications, providing full acknowledgment of user performance quality indicators (Table 6.7).

6.3.3.4 Protocol Analyzer

Protocol analyzers allow to capture and examine packet data from the network. Depending on the interface the analyzer is connected to, it may capture the information about different layers in the network. Typically, protocol analyzers are used to monitor TCP/IP packets and application level information (HTTP, FTP, etc.) in the client side, and they run on a computer connected to the wireless device. For this kind of application, the wireless connection works just as a bearer (which is totally transparent), and only throughput, delay and packet losses are affected. Thus, this kind of tool is totally technology independent.

Protocols analyzers can be also used in the server interface, for example when the server is under the control of the team doing the test, it allows to better understand the end-to-end dynamics of the communication.

Additionally, inside the wireless network, protocol analyzers may be used to 'sniff' several internal interfaces, which may allow to detect specific problems and sources of delays. Typically,

Table 6.7   Main KPI per service used in driving testing

Service	Main KPIs
PING	Delay of isolated ping
	Average delays of pings
FTP	Instantaneous and average throughput
Web browsing	Download time per Web page
	Size and number of objects in the Web page
	Instantaneous and average throughput during Web download
Streaming	Time evolution of received frames/s
	Number of re-buffering events
	Real-time audition and visualization of streaming files
WAP	Download time per WAP page
	Size and number of objects in the WAP page
	Instantaneous and average throughput during WAP download
PoC	STS delay time and VDT

Gn, Gi or Gp interfaces can be monitored with standard TCP/IP-based protocol analyzers. Other interfaces like Abis or Gb in GPRS network require more specific tools.

Of course, analyzing in detail the TCP traces corresponding to every executed test cases is a time-consuming task that would add little value to normal reporting process. However, it is really important to take TCP traces systematically at least during driving tests, in order to troubleshoot some cases if needed afterward.

There are many commercial applications in the market, including post-processing capabilities which may provide additional facilities to the analysis, but there are also free, very well-known and powerful applications such as Ethereal[©] [13] which provides reasonable easy-to-use capabilities (shown in Figure 6.7).

6.3.3.5 Network Interface Analyzer

There exist tools similar to TCP/IP protocol analyzers but specifically designed for internal interfaces in the wireless network. They normally provide full trace of all packets transmitted through the observed interface, for example, Gb, Abis, etc.

However, due to the complexity and the great amount of data that this kind of tools generate, it is of great importance to count also with proper post-processing applications that minimize the analysis time and reduce the number of post tasks to be done by the engineers.

(a)

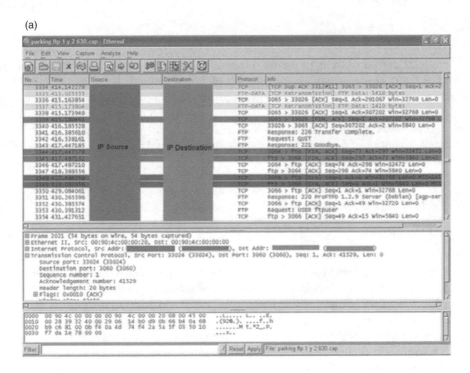

Figure 6.7   Example of Ethereal[©] TCP protocol stream. (a) Message flow. (b) Graphical representation

(b)

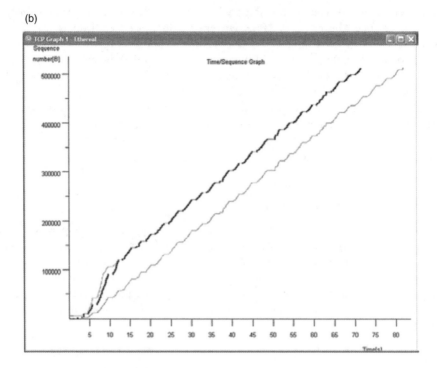

Figure 6.7   (*continued*)

These tools are normally used during testing of new features or equipment, and during network roll out. However, although the information that they provide is very detailed and can be very useful, they require the physical location of engineers close to where wires and software are deployed, and increases the complexity of the trial planning. For this reason, the usage of this kind of tools is normally limited to specific cases where there is a justification for a detailed analysis of the interfaces, like troubleshooting and delay analysis. They are also typically present in lab environments.

6.4 Technology Benchmarking

6.4.1 Introduction

This section presents a performance benchmarking of different radio technologies. Note that this audit has been fulfilled over different commercial mobile networks over which there was not any control, focusing on application level performance (with no possibility to obtain and relate with radio performance metrics) and with limited resources to perform an extensive measurements campaign.

In such a situation, regular data terminals, laptops and standard applications to test the different services as well as regular service plans provided by the different operators were used. The multi-technology scope of the audit (GPRS, EGPRS, WCDMA, CDMA2K-1x) required the work in different locations worldwide, being difficult to assess the performance of the different technologies under similar conditions. In addition, the scope of the tests needed to be restricted to a user level and a restricted number of services. Specially considering that budget limitations would make that in most of the cases, the voluntary testers would had to be also the owners of the equipment (and even pay for the service).

Thus, the limitations established by the test restrictions must be taken into account when looking into the results, and consider that all the measurements, although repeated several times in order to minimize statistical errors, only cover few specific locations in different networks. Therefore, these results cannot be considered as a final proof of the performance of each technology. It must be noted that the radio quality of each test may vary, although it was always tried to obtain good-quality locations (i.e. maximum coverage), and that the evolution status of the operators may vary: parameter optimization, equipment manufacturer, hardware/ software versions, etc.

Nevertheless, the results obtained in such conditions pretend to provide a general view of the real range of performance that anyone may obtain from networks which are already up and running, as well as to show up most common interactions between the radio capabilities of the systems and the measured end-user performance.

In addition to the practical measurements, this section will also cover basic concepts of traffic generation and some aspects about the planning of the tests.

6.4.2 Traffic Generation

The knowledge about the way the traffic load behaves in a network is a key issue at the time of dimensioning, optimizing and testing a mobile network. This behavior is often described by means of models, which can be understood as mathematical abstractions with different levels of complexity that try to imitate one or several statistical characteristics of a real type of traffic or a specific traffic flow.

Before the first computer networks came up, the well-known Poisson model was enough to describe the call arrival process of the telephony users and the exponential distribution of the duration of their calls, which were used to dimension these circuit-switched networks with a blocking probability according to the Erlang-B formula.

Today, data networks are experiencing a huge and rapid growth, and the variety of services that can be offered within these networks is wider day by day. In circuit-switched networks all the users seem to have the same behavior concerning traffic load generation whereas in packet switched networks the generated load is very dependent on the service the user is accessing. For example, audio-streaming services may require a constant bit rate and other services like WAP have a very bursty nature. In addition, the Poisson model has been proven to be incapable of characterizing these new kinds of traffic. Because of that, there is a need for the definition of more accurate models that fit better with the real traffic characteristics, so that data networks can be simulated, dimensioned, optimized and tested.

The creation of a model starts with the analysis of traffic samples taken in real networks or available in databases. After that, the time scale for which the model will be applied must be considered. It is clear that a more accurate model can be built if we concentrate on a particular time scale and avoid effects that are rather not of interest to us. For example, if we wanted to build a model of the traffic generated during the busy hour in a cellular network, perhaps we would not be concerned about how the traffic load could increase in the next two years. This means that the more we delimit what we want to model, the better the model will be. Finally, the purpose that the model was meant to has to be considered, as for example we may be interested in modeling the traffic generated in a network at application layer or may rather want to include specific procedures at the lower layers of the protocol stack.

As mentioned before, the quality of a model depends on how it fits to the real traffic. But this is not the only point to be considered at the time of choosing a model. There are other factors like its mathematical complexity, ease implementation, number of parameters, capacity to model other kinds of traffic or just its own aggregated traffic [4]. For example, we may not want to use a very accurate model in a simulation tool if it takes a lot of time to be implemented or if it is highly CPU demanding.

Traditionally, there are two approaches to build a traffic model.

1. *Behaviorist* approach: The usage of mathematical models with a certain number of statistical parameters does not take into account the causes that generate the real traffic. These models are based on counting the number of generated packets during a specific period of time and do not include anything about the call arrival process.
2. *Structural* or *hierarchical* approach: The model is based on the underlying structure of the system that generates that traffic, so that the parameters of this model can have a physical meaning. This feature makes structural models suitable to follow the evolution of the traffic load, the user behavior, the number and size of the transfers, etc.

There are uncorrelated 'behaviorist' models that generate every sample by using the same statistical distribution function with no dependence between a sample and the previous ones. These models have been proven to be useful to imitate certain aspects of traffic generation, like the call duration and the size of transferred files. However, as data traffic is usually bursty, it is necessary to model certain correlation between samples in the traffic generation model. This is the reason why the Short Range Dependencies (SRD) models are introduced. SRD models could be considered as a special type of 'behaviorist' models, as the underlying structure of the entity that generates traffic is not taken into account in any case. Examples of this type of model are Markov-chains-based processes and traffic models that use Autoregressive Moving Average (ARMA) filters.

But the Internet traffic has proven to have Long Range Dependencies (LRD) that cannot be modeled with uncorrelated or SRD models. Several researches in [6–8] have shown that the Internet traffic generated during a specific period of time is dependent on the traffic generated in previous periods of time and that the Internet traffic is self-similar, which means that it always follows the same statistical distribution independently on the length of the time interval in which the account of packets takes place. This strange phenomenon has been often explained by falling back on multilevel traffic generation structures. These structures are just the base of the structural models, which try to reproduce the way traffic is generated by imitating the underlying structure of the system that generates traffic, as mentioned before. A research about how HTTP traffic is characterized with structural models can be found in [5]. Figure 6.8 shows an example of how the FTP traffic generation can be modeled by a multilevel structure.

The model depicted in Figure 6.8 consists of two levels: session and transfer levels. The first level is the session level, whose role is to model the creation and duration of an FTP session. An FTP session encompasses the time since the FTP client is started until it is finished. On the other hand, the transfer level corresponds to the transfer of a file or any other kind of data sent through the FTP data connection. In our example, the model is defined by using five parameters that can follow any kind of statistical distribution. For the session level, the time between the beginning of sessions and the number of transfers per session are considered,

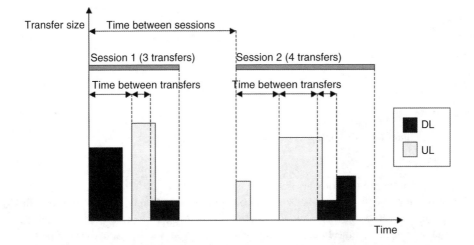

Figure 6.8   Example of FTP multilevel model

whereas for the transfer level, the time between the beginning of transfers, the probability of downlink (DL) or uplink (UL) transfer and the sizes of the transfers are considered.

6.4.3 Test Case Definition for Benchmarking

In a live testing with limited time and resources, amount of tests that can be performed is very limited, and cannot by any means follow any complex distribution unless a detailed load testing is pursued, but this case is typically more in the scope of simulations or controlled lab testing.

In our case, the tests were limited to several typical applications as already described in previous chapters. The tests would be performed in sequence during low load periods (i.e. early in the morning or night time), by an application which allows to define the whole sequence of test, and automatically collects all the application and service level KPIs, which would reduce the work to be made by the person carrying the test to configure the data connection and just monitor that the test was executed properly. Additionally, for every test, a protocol analyzer was activated in order to capture the TCP/IP packets for further analysis.

The following were the selected services.

- *Ping*: With the aim to measure the network RTT. Different packet sizes were selected to analyze the impact of transmission rate and establishment delays. Different delay separations between pings (0, 1 and 10 seconds) were selected in order to study the effect of channel establishments.
- *FTP*: With the aim to measure the throughput of the different applications.
- *HTTP*: Provides the behavior of interactive service and impact of request–response delays.
- *Streaming*: Provides performance measurement of real-time service experience. Quality and delay are checked for different video qualities.

All the test cases were defined in the form of text scripts that were to be interpreted by the measurement tool, as described in Figure 6.9:

```
#connect 1
#Streaming view
streaming rtsp://IP Address/intacto_5.00_15_160x120_MPEG4.3gp  200
streaming rtsp://IP Address/intacto_10.00_15_160x120_MPEG4.3gp  200
streaming rtsp://IP Address/intacto_15.00_15_160x120_MPEG4.3gp  200
streaming rtsp://IP Address/intacto_25.00_15_160x120_MPEG4.3gp  200
streaming rtsp://IP Address/intacto_35.00_15_160x120_MPEG4.3gp  200
#discon ect
```

```
connect 1
#Web download
loop 5
web IP Address/google.htm
web IP Address/wanadoo.htm
web IP Address/yahoo.htm
web IP Address/elmundo.htm
web IP Address/thenewyorktimes.htm
end_loop
disconnect
```

Figure 6.9   Example of test script description for streaming and Web

In order to fully control the contents that were used in the tests, different application servers for FTP, streaming and Web were activated in our premises in Malaga, where the test files and Web pages were to be stored. However, the fact of having one common server for all the tests has the drawback that the Internet component of the delays suffered by tests done over the sea would typically be higher than the ones made in the same country.[1] However, we had to live with this effect and consider it in the analysis.

6.4.3.1 Test Location Selection

Test location was very much determined by two factors. Availability of technology and contact person in the area, with the only condition that all the cases should try to find good coverage conditions (i.e. maximum number of coverage bars in the terminal), so as to obtain comparison of the networks under best radio conditions. Therefore, for the GPRS and WCDMA, the tests were to be done in Europe, specifically in our office and home in Malaga (Spain), while EGPRS and CDMA2000-1x took place in different cities in the US (Atlanta, Las Vegas, Miami, Philadelphia) and South America (Lima, Quito, Santiago) with the support of several colleagues who were located or in business trip in those areas (Figure 6.10).

Most of the locations were indoor but with good coverage, although in some cases, looking for reasonable good coverage, some tests were also done inside a car in a parking lot.

6.4.4 Benchmarking Result Analysis

As described previously, the objective of this section provides a service performance benchmarking over the most important radio access technologies. The goal is to provide typical performance measurements for a set of data services and show a performance comparison between technologies. The performance results are based on real measurements taken from GPRS, EGPRS, UMTS and CDMA1x cellular networks.

---

[1] Actually, this is not necessarily true, and depends on the routing path from the cellular operator's network toward the fixed line supporting the server connection.

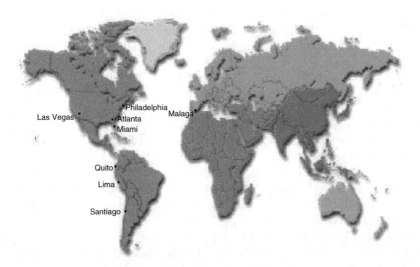

Figure 6.10 Location selected for the tests

Table 6.8 Terminal capabilities

Terminal technology	Capabilities
GPRS	3 + 1/2 + 2 TSLs (multislot class 6)
EGPRS	4 + 1 (multislot class 8)
WCDMA	384 kbps[2]
CDMA2000-1x	153.6 kbps

The analysis will be performed for each service separately, but considering the special characteristics of each of them. Capabilities of the terminals used in the different tests are included in Table 6.8.

Application severs used for the tests were located in the Internet, not inside the operator's network as it is typically done for testing purposes. This is something to be taken into account due to the random behavior that the Internet may add to the results. The main KPIs of Ping, FTP, HTTP and streaming services are analyzed over the different technologies.

6.4.4.1 Round-Trip Time

The assessment of the RTT is one of the first steps during a service performance analysis process. Already in Chapter 5, the huge importance of network latency on the overall service performance was described.

---

[2] Note that during the analysis it was found that UL throughput for WCDMA was limited to 64 kbps, although this limitation is expected to be due to network design decision rather than terminal limitation.

This section focuses on assessing the RTT of different networks by using ping commands. Ping performance results correspond to the delay of sending a configurable packet size in the UL direction to a server until the same packet arrives back to the terminal. Note that both UL and DL data rate capabilities are affecting considerably the results, mainly for big packet sizes. Also note that these values are slightly different to the typical RTT suffered during a TCP-based connection, where commonly, a big packet is sent in one direction and a small packet (ACK) is sent back in the reverse direction.

We will also consider the time between consecutive ping requests in order to evaluate the difference between sending several consecutive ping requests and sending just one isolated ping request. The time between consecutive requests was set to 0, 1 and 10 seconds. This separation is the time elapsed from the moment a request is fulfilled until a new request is sent.

Figure 6.11 depicts the average RTT for ping requests, where different number of samples were considered (>100 in different locations), with a different packet size in GPRS, EGPRS, UMTS and CDMA1x networks.

Starting with GPRS and EGPRS technologies, measurement results for ping requests of 32, 256 and 536 bytes show higher delays for EGPRS than for GPRS, in contrast to ping requests of 1500 bytes, where EGPRS delays are lower. The fact that small ping requests experience higher delays in EGPRS is due to a partial deployment of EDGE. Since the PBCCH (Packet Broadcast Common Control Channel) and EGPRS Channel Packet Request are not available in this network, the TBF establishment has to be done in two phases (TWO-PHASE access) for EGPRS, which means an extra delay equivalent to 1 RTT in the radio interface [11]. However, for GPRS it is possible to establish a TBF in one phase

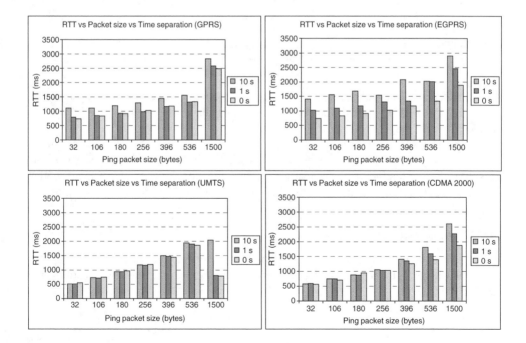

Figure 6.11   Ping performance in GPRS, EGPRS, UMTS and CDMA1x networks

(ONE-PHASE access) leading to lower delays. The extra delay in EGPRS because of the TWO-PHASE access is compensated by a higher air interface capacity, especially when ping requests of 1500 bytes are sent; for big packet sizes the throughput of the radio interface has a great impact on the total delay. This is the reason why the average RTT for ping requests of 1500 bytes is lower in EGPRS than in GPRS.

The GPRS DL TBF establishment delay can be derived from these figures. For ping request separated by 10 seconds, new TBFs have to be established both in UL and DL. However, for ping request separated by 1 or 0 second, only one UL TBF has to be established, since the delayed DL TBF release feature is active [16]. Therefore, the DL TBF establishment delay can be estimated as the difference between the RTT of requests separated by 10 seconds and requests separated by 0 second. This procedure returns a value of 290 ms approximately.

Regarding the UMTS figures, a uniform increase of the RTT can be observed as the packet size grows, except for the case of packets of 1500 bytes when consecutive ping requests are separated by 0 and 1 second. This effect is due to the DCH (Dedicated Channel) establishment. As explained in Chapter 5, RRC CELL_FACH state is used until a DCH is established (RRC CELL_DCH state) when a configurable traffic volume threshold is exceeded. It can be derived from the figure that a DCH is established for the first request of 1500 bytes (so the threshold in this particular network is between 536 and 1500 bytes), which means an extra delay for that request (typically around 1 second); but subsequent requests can experience a lower delay because of the higher data rates this DCH provides.

Ping requests separated by 10 seconds do not experience less delay because the separations of 10 seconds between samples are more than enough for a DCH to be released (typically around 1 or 2 seconds). This way, for every request of 1500 bytes with a separation of 10 seconds, a new DCH is always established and released. Therefore, it is possible to estimate the DCH establishment delay as the difference between the RTT of the requests of 1500 bytes separated by 10 seconds and requests of 1500 bytes separated by 0 or 1 second. According to the results, DCH establishment delay takes 1.2 seconds approximately.

The throughput of the FACH can also be derived from the Figure 6.11 by considering how much the RTT increases as the packet size grows. For example, computing the requests of 536 bytes and 396 bytes, the calculation would be as follows.

$$FACH\_throughput(kbps) = \frac{(536 - 396)bytes * 8\ bits/byte}{(RTT(536\ bytes) - RTT(396\ bytes))(ms/2)} \cong 5.2 kbps$$

In the case of CDMA1x, a similar effect than the one just commented for WCDMA is observed for 536 and 1500 byte pings. In these cases, the delay observed for 10 seconds separation between samples is significantly higher than for the 1 and 0 second separation, indicating a faster channel establishment or reuse of previously assigned bearer in these cases.

It is also worth comparing network latencies between technologies. Apart from the (E)GPRS discrepancies due to the TBF establishments commented previously, the reader should note the huge improvement of network latency that UMTS and CDMA2K networks provide compared to (E)GPRS networks. This latency reduction (up to 50% in some cases) will have an important impact on the performance of the rest of the services (mainly those based on TCP). The main reason for these different latencies may come from network architecture and specific protocol requirements in order to assign a channel.

6.4.4.2 FTP Performance

This section tries to compare the throughput of the different radio technologies (GPRS, EGPRS, UMTS and CDMA1x) by transferring files of 100, 300 and 500 kB via FTP. Figure 6.12 depicts the results obtained.

As expected, this figure shows how FTP throughput increases as the file size grows, regardless of the cellular technology. This behavior is mainly due to TCP effects like the slow start and the three-way handshake during TCP connection establishment. The bigger the file transferred, the lower the weight of these TCP effects on the overall transfer delay. However, the trend that throughput follows for the different radio technologies varies, and has a lot to do with the RTT. As commented in previous sections, the extra delay that both the slow start and the three-way handshake introduce depends on the RTT and throughput.

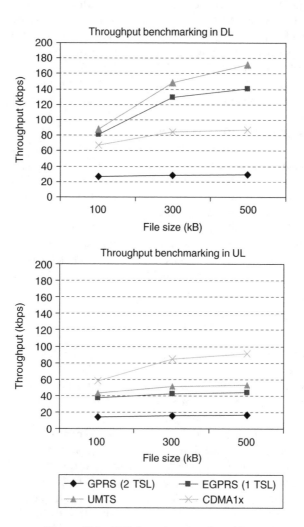

Figure 6.12　FTP throughput benchmarking

It is also worth commenting that the impact of these TCP effects on the average throughput is less important as the maximum achievable throughput decreases. This is due to the fact that the higher the maximum throughput, the longer the time elapsed in the slow start phase. For example, although in GPRS the RTT is higher than in UMTS, there is not much difference in the throughput achieved in GPRS for files of 100, 300 and 500 kB because the maximum achievable throughput does not exceed 30 kbps. In a nutshell, impact of slow start is higher the higher the BDP.

In Figure 6.13 it can be seen the different behavior from TCP/IP point of view of EGPRS and CDMA FTP transfers, and which can explain why CDMA tends to show better values than expected while EGPRS sometimes tends to show lower values.

Let us look first at the DL transmission. The first difference is the duration of slow start phase, which is up to 8.5 seconds for EGPRS while only 4.9 seconds for CDMA. At this phase, the throughput is limited by the TCP layers instead of the radio link layer (approximately 57 kbps for EGPRS and 47 kbps for CDMA). Note that in the case of EGPRS, for a 512 kB file this means up to 30% of the transmission time, while for CDMA the slow start is roughly a 10%. On the other hand, the transmission seems to be done in chunks of data which introduce a small delay which is proportional to the RTT (higher in EGPRS). Note that for some reason, the size of these chunks is smaller in EGPRS (thus multiplying the effect). One possible explanation might be that EGPRS connection is using default configurations from Windows XP dial-ups, while the CDMA card includes a software which controls the dial-up and might be trying to optimize the performance of TCP.[3]

In the case of UL, similar behavior is observed, the only difference is that now the slow start time of the EGPRS connection is reduced to approximately 7 seconds, and the separation between the 'data chunks' is more appreciable (mainly because of the different scale of x-axis).

However, even with this analysis the good performance of CDMA in UL compared with EGPRS and even WCDMA seems difficult to explain. But, let us first comment the assumption for CDMA, which is that the network configuration is allowing symmetric throughput in UL and DL as far as the interference level allows it and the phone (in this case PCMCIA card) supports it. This assumption seems quite reasonable, as the measured values in both directions are almost the same, and the CDMA network where the tests were done is already in a mature state.

On the other hand, it is a design option in WCDMA networks choosing to support only 64 kbps in the UL if that configuration fits with expected services in the network (i.e. even videoconference is possible with 64 kbps). This actually may allow cheaper deployments, specially for young networks. It is our assumption that this design is in place, as the 64 kbps at radio level fits quite well with the roughly 60 kbps in average obtained at application level.

Finally, EGPRS is using only 1 TSL in UL direction (due to limitation of the terminal), and thus the available radio throughput is only ¼ of the DL capacity. Note that the effective factor UL/DL at application throughput is ⅓ instead of ¼. The reason is again the impact of the slow start, which will be bigger as the BDP increases. In the case of UL

---

[3] The software coming with the CDMA card also included a data compression application, although it was not installed for the purpose of these tests.

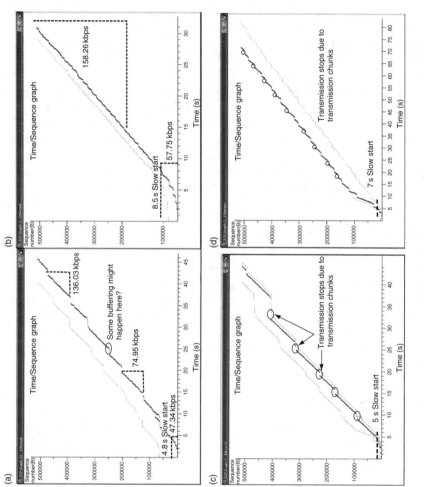

Figure 6.13 EDGE vs CDMA FTP+TCP/IP behavior. (a) CDMA[4] (b) EGPRS (c) Download (d) Upload

[4] Note the evolution of throughput allocated to CDMA channel (Supplemental Channel) follows theoretical trend as described in Figure 2.18.

with 1 TSL, the BDP is also reduced to 1/4, being the RTT constant compared with the DL case (4 TSL).

6.4.4.3 HTTP Performance

This section analyzes the performance of HTTP, the most common service in the Internet, when it is accessed from a cellular network. A reduced group of standard Web pages, corresponding well-known Internet sites (search engine, newspaper, leisure, etc.) were selected and stored in Private server, in order to ensure that all the tests were downloading exactly the same information.

As it was described in Chapter 3, Web pages are composed of HTML text which may link to one or several objects (e.g. images, JavaScripts, Flash, etc.). In general, the response time when downloading a Web page will highly influence on the structure of the page. The main reason comes from the inherent HTTP implementation in current browsers, with most of them supporting HTTP 1.1 without *pipelining* feature. Although this version of HTTP allows reusing TCP connections to retrieve subsequent objects, the different requests within the same TCP connection cannot be sent in parallel (i.e. the request of a new object cannot be sent until the previous object has been received). However, more than one TCP connection may be activated in parallel (normally 2).

The list of Web pages used for the tests as well as the main characterization parameters are described in Table 6.9 and Figure 6.14.

Figure 6.15 illustrates the performance (in terms of response time) for a Web browsing service when downloading the list of Web pages described in Table 6.9 over a GPRS, EGPRS, UMTS and CDMA1x networks.

The results show important performance degradation as the number of objects included in the Web page increases. For instance, the response time of a 105.94-kB Web page with

Table 6.9 Sizes, URL and number of objects of the downloaded Web pages

URL	Total size (kB)	Text size (kB)	Number of objects
http://www.google.es	14.87	4 632	5
http://www.wanadoo.es	80.00	65 866	32
http://www.yahoo.es	105.94	81 186	18
http://www.elmundo.es	272.35	97 791	79
http://www.thenewyorktimes.com	285.34	77 066	84

Figure 6.14 Web pages used for the tests

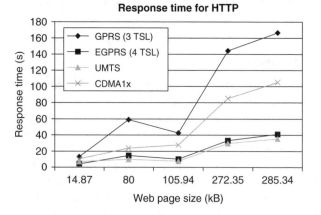

Figure 6.15   HTTP response time in GPRS, EGPRS, UMTS and CDMA1x networks

18 objects is lower (in most of the cases) than the response time of a 80-kB Web page with 32 objects. In the last case, the bandwidth utilization is reduced by the numerous HTTP request messages.

Additionally, higher burstiness of the traffic may lead to a worse performance at the radio interface due to different radio resource management procedures. For instance, in (E)GPRS networks a higher number of TBF establishments and releases will occur (adding additional delays). Longer RTTs in the network will cause higher performance degradation.

Apart from this effect, it is important to note the higher response time difference among radio technologies as the Web page size increases. As expected, both data rate capabilities and network latencies influence on the overall result. As shown in previous sections, higher bit rates are achieved with EGPRS and UMTS terminal, reducing considerably the response time. However, the lower UMTS latencies (compared to EGPRS) make this technology perform a bit better even though the available UMTS bearer is a bit lower in these tests.

These results also show the big impact of application layer in the service performance, introduced not only by the application messages but also by how HTTP manages the lower layer, i.e. TCP connections. For instance, comparing HTTP vs FTP results (from previous section) for similar content sizes, up to 50% throughput reduction due to application layer is introduced.

6.4.4.4 Streaming Performance

This section describes the streaming service performance when it is accessed from GPRS, EGPRS, CDMA1x and UMTS networks. QuickTime 6.5 streaming client as well as Darwin Streaming Server have been used in the tests [2, 3].

For this measurement campaign, several video streams with different bit rate requirements were prepared. These streamed media are usually referred to as stored media or video on demand. The main characteristics of these test video streams are summarized in Table 6.10.

The results presented in this section correspond to real-time streaming. This type of streaming uses RTP/RTSP and is the one required for live feeds like TV channels, live events

Table 6.10 Characteristics of the video streams

Bit rate	Frames (per second)	Size (pixels)	Format	Duration (seconds)
5.0	15.0	160×120	MPEG4	105
10.0	15.0	160×120	MPEG4	105
15.0	15.0	160×120	MPEG4	105
25.0	15.0	160×120	MPEG4	105
35.0	15.0	160×120	MPEG4	105

and delivery of long-form video. Among other advantages, real-time streaming allows random access within prerecorded movies, the movie is not stored in the client's hard drive and only the bandwidth needed is used. In real-time streaming the user is supposed to display the media approximately at the same time they are being generated, in contrast to progressive download (also known as fast start).

In real-time streaming, the client always tries to synchronize with the source of the stream. Therefore, concepts like the number of re-bufferings and their duration are not quality indicators for this service, as they are in fast start streaming. In contrast to fast start streaming, in which the media are displayed with the same quality they were coded with and the playback can be stopped until there is enough information to resume it, in real-time streaming the media can be displayed with lower quality if the data rates exceed the connection speed. This way, the main KPI for this service is the quality of the video stream, i.e. how similar what the user is experiencing is to the original media.

Therefore, the quality of the video stream is evaluated as the average number of displayed frames per second (Figure 6.16) over the original number of frames per second of the video stream in percentage, since the frame rate relates directly to the smoothness of the playback

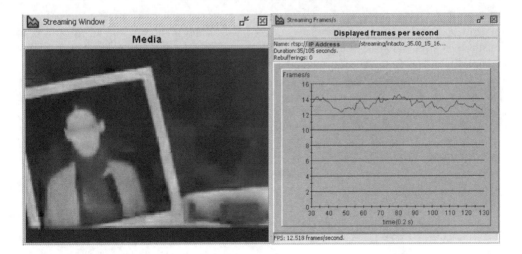

Figure 6.16 Example of streaming test: Image + Frames/s time evolution

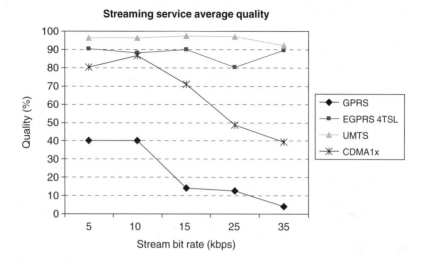

Figure 6.17 Quality of the streaming service

the user perceives. Figure 6.17 shows the quality for a real-time streaming service when the media in Table 6.10 are accessed through GPRS, EGPRS, UMTS and CDMA1x networks.

As expected, the quality of the streaming service decreases as the required bit rate increases. This effect is not clearly experienced in UMTS, since the required bit rate of the most demanding video stream is easily achievable in this technology. Note that there was no QoS negotiation before taking these measurements.

6.4.5 Network Performance with User Multiplexing

This section provides some real performance measurements intended to show how the network performance and end-user experience are degraded as the number of users grow. These results have been obtained from FTP and HTTP services in a real GPRS network. Due to the complexity of the test, number of required terminals, etc. it was not possible to perform similar analysis for different technologies, but the results shown here mean to describe the high impact that resource sharing will have on any cellular network, specially when comparing with ideal throughput values.

In this test, a multi-terminal traffic generator [9] was used to generate traffic through a number of mobile phones that were used as modems by a computer. This software emulates the activity of mobile users by characterizing their behavior with probability distribution functions and driving the resulting traffic through multiple terminals. The connections between the computer and all mobile phones are established via Bluetooth technology, as shown in Figure 6.18.

Measurement results have been taken at application level, where FTP and HTTP services were tested. Since FTP is a very bandwidth demanding service, it can provide a clear picture of how radio interface capability is being used and shared. Several FTP downloads have been performed for a number of mobile phones ranging from 1 to 7. For the realization of the

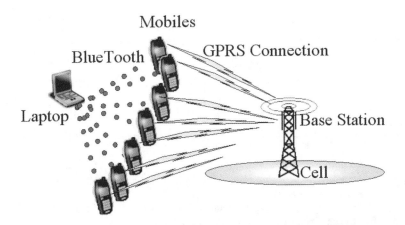

Figure 6.18   Traffic generation with multiple mobiles used as modems

whole array of tests, the FTP and HTTP servers were located in the Internet, and not in the operator's network. This means that the random behavior introduced by the Internet should also be considered. GPRS terminals with $4+1$ TSL capability were used.

It must be taken into account that no cell barring was performed in the cell under test. Nevertheless, although the activity of other users in the same cell could affect some of the results obtained, due to the time at which the tests were done, and due to the low utilization of the GPRS service in the test area we assumed that there were no other users interfering with the measurements.

6.4.5.1 FTP Performance

The impact of introducing new users in the system on the average FTP performance is analyzed in this section. In order to carry out this analysis, two kinds of experiments are done. In the first one, FTP performance is studied when all the users are downloading files consecutively, whereas in the second one they are downloading files on a probability distribution function basis.

FTP performance is computed by averaging the throughput obtained in all the transfers of the users. In the test for consecutive FTP downloads, every user is continuously downloading a 300-kB file repeatedly until the end of the test, i.e. during 7 minutes. This way, every terminal is generating as much traffic as it can.

Figure 6.19 shows the degradation of the application throughput as the number of users increase. This figure also depicts the success rate of the transfers, which could be understood as the percentage of successful transfers, i.e. how many of the transfers were completed in percentage. This measurement is a key indicator of how the network performs from the end-user point of view, since it has a lot to do with his experience.

Figure 6.19a shows the average FTP throughput of the GPRS terminals acting as a modem for a Linux machine. In principle, the figure seems to show an unexpected behavior in the performance since the throughput per user curve is increasing from 3 to 6 users sharing the resources, but the reason comes from the success rate results. Note that throughput values in

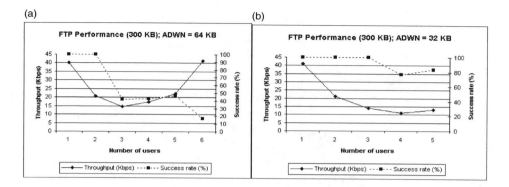

Figure 6.19 FTP Average throughput vs number of users (a) Default Linux Red Hat 9 advertised window (64 kB); (b) Decreased advertised window (32 kB) [9]

the figure are only computed for those FTP sessions that has finalized correctly. As many transfers are aborted when a considerable number of users are sharing the resources, the transfers that persist can take advantage of the silence periods of those users that have to establish a new FTP session (data connection establishment, slow start, etc.). Going into the details of this test, it was possible to verify that the root of this problem was the default TCP configuration of the receiver (Linux RedHat 9 machine), which defines a default TCP advertised window of 64 kB.[5] As explained in section 5.4, advertised windows much higher that the Bandwidth Delay Product (BDP) may lead to performance degradation, since the sender is congesting the network and the acknowledgments may arrive back too late. This is exactly the case with GPRS, for which BDP is normally around 4 or 5 kB. In that situation, the delay in possible retransmissions increase hugely and hence, TCP socket timeouts may produce session releases.

By analyzing the TCP trace, it was possible to check that the TCP advertised window was around 64 kB, which means that it would take up to 49 seconds to transmit this window when the throughput is around 10.4 kbps, as it was measured at application level. Those 49 seconds were proven to be more than enough to make the TCP retransmission timer expire several times and finally drop the TCP connection. Because of this, it was decided to reduce the TCP advertised window to 32 kB as its maximum value. The results with this new TCP configuration are depicted in Figure 6.19b.

Figure 6.19b shows the same tests but using a new TCP configuration (advertised window = 32 kB). The success rate with this new configuration is improved considerably and resource sharing is more equitable among users, i.e. the average user throughput decreases as K/N approximately, where K is the total throughput capacity of the radio interface and N is the number of users (cell under test supports 4 GPRS TSLs).

---

[5] Note that 64 kB is a typical default value for current OS like Windows XP and Linux. However, older versions like Windows 2000 or Windows 98 used to have 32 kB or even 8 kB as default values.

6.4.5.2 FTP Performance with Statistical Multiplexing

In order to evaluate the statistical multiplexing in a cellular network, an FTP traffic source is defined according to a given number of Erlangs, so that the FTP performance evolution with the number of Erlangs can be analyzed. As the traffic generation tool defines an FTP traffic source by means of probability distribution functions for the size of the file and the time between consecutive downloads, a transformation from Erlangs to these functions is needed. Seven mobile phones were used to send the data traffic coming from this source.

The exponential probability distribution function was chosen for both the size of the file and the time between consecutive downloads. This function is represented by the following formula.

$$f(x) = \lambda e^{-\lambda x} \text{ for } x \geq 0$$

where 'x' can be the size of the file in kB or the time between downloads in seconds, and λ can represent the inverse of the average size of the file in kB^{-1} or the average number of downloads per second (s^{-1}).

If the average size of the file (AvFile) is set to 500 kB and a TSL capacity of 10 kbps is assumed for GPRS (i.e. 1 data Erlang is equivalent to 10 kbps), then the average time between consecutive downloads (AvTime) is given by the following expression.

$$AvTime(s) = \frac{AvFile(kB) \times 8 \frac{bits}{byte}}{10 \text{ kbps} \times Data\_Erlangs} = \frac{400}{Data\_Erlangs}$$

Measurements were taken for 1–5 data Erlangs, where results for a given number of data Erlangs were always taken starting at the same time of the day for 5 hours. Figure 6.20 shows the evolution of the FTP throughput with the number of data Erlangs. Note that these results may present certain deviation due to other circuit-switched and packet-switched users (since the cell under test was not barred).

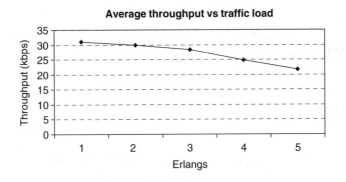

Figure 6.20   Average throughput vs Data Erlangs

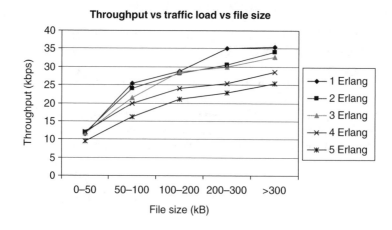

Figure 6.21   Average throughput vs data Erlangs vs file size

Because of a softer resource sharing (i.e. a statistical multiplexing), the throughput does not decrease in this case as fast as it did in the measurements of the previous section.

In Figure 6.20, the effect of downloading files with different sizes is not shown, as the throughput is directly calculated on transmitted data of partial transfers without considering establishment delays. The size of the downloaded file is very important when throughput measurements are taken, since typical TCP effects like the slow start or the three-way handshake may reduce its value. Figure 6.21 depicts how the FTP throughput evolves for a number of data Erlangs ranging from 1 to 5 and different intervals of file size.

Figure 6.21 shows that when the size of the downloaded file is small (<50 kB), the average throughput does not depend much on the traffic load. This makes sense because when the downloaded file is small, a very important component of the transfer delay is due to TCP effects like slow start, three-way handshake, retransmissions, etc. and the radio interface capacity does not play a key role. On the other hand, when the size of the file is big (>100 kB), the biggest percentage of the transfer delay corresponds to the radio capacity and the achieved throughput depends a lot on the traffic load, as that radio capacity is shared between users.

6.4.5.3 HTTP Performance

This section analyses the HTTP performance when the number of users increase in the cellular network. HTTP performance is computed by averaging the throughput obtained for all the transfers. Every user downloads a Web page of 58.5 kB (consisting of 22 objects) repeatedly until the end of the test, i.e. during 3 minutes, and the TCP advertised window was set to 32 kB.

Figure 6.22 shows the average user throughput evolution when the number of users increase for 3 and 6 TCP connections to download the embedded objects.

As expected, the HTTP performance decreases as the number of users in the system increase. But it should be noticed that this decrease is different from the one experienced in

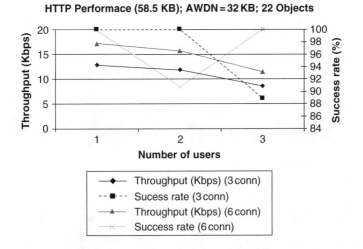

Figure 6.22 HTTP average throughput vs number of users

FTP, as seen in previous section. This fact is explained if we consider that the HTTP traffic is more bursty than FTP (e.g. HTTP needs to open several TCP connections to download objects and has a higher application layer overhead) which means that there are periods of time in which some users are transmitting at low rates. These periods can be used by other users to transmit at higher rates, which lead to a softer resource sharing known as statistic multiplexing.

Figure 6.22 also shows a better performance when using 6 TCP connections per user instead of 3. A higher number of TCP connections allows to minimize the effect of slow start since several connections are used simultaneously. Note, however, that having too many TCP connections in parallel may also have some adverse impact due to additional overheads, and specially because the number of TCP connections of servers and clients is normally limited, which might cause a situation of 'TCP connections shortage' if there are many concurrent applications connected to the Web.

6.4.6 Push-to-Talk over Cellular (PoC)

Figure 6.23 shows a PoC performance comparison of the main KPIs of this service: time from pressing the PTT button until the sonorous beep of the floor granted arrives (STS), and the time when the speaker starts talking until the receiver starts listening to the voice (VDT). The measurements have been performed in GPRS, EDGE and iDEN networks. PoC is a novel service in (E)GPRS networks whereas it is a mature application for iDEN networks. This issue has to be considered when analyzing values in Figure 6.23. Measurements related to iDEN were taken in a live network where the service was deployed long time ago. On the contrary, GPRS and EGPRS measurements have been taken using different providers all them still trialing their systems. The high standard deviation of GPRS and EGPRS values in Figure 6.23 reflects aforementioned trialing status, tests that have been carried out over different networks and with different vendor solutions. Some of the solutions, probably the most mature ones, provided much better performance, however the average delay is higher due

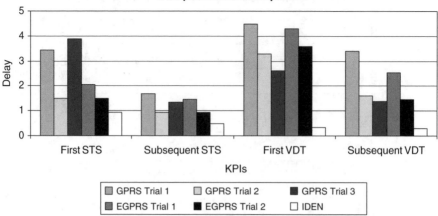

Figure 6.23   PoC performance comparison

to the weight of worst solutions. Therefore GPRS and EGPRS provided values that have to be considered only as a reference of possible values based on the current situation and not as the optimum performance value that can be reached with this technology.

For both KPIs, STS and VDT, there are different values taken into account depending on whether it is the first time the user is accessing to the system, after a long inactivity period or not. In all the cases, the first access has longer delays, but based on the implementations the difference between the first and the subsequent access can be higher or lower, i.e. for (E)GPRS an implementation can try to keep the TBFs established during the PoC session aiming at decreasing the delay of the subsequent access.

In general, EGPRS provides shorter delays than GPRS as its transmission rate is higher. However, still at the moment of performing these tests, EGPRS needs more time in the TBF establishment due to the two-phase access, which leads to longer delays than in GPRS. However, faster access for EGPRS is already available for a number of systems. The different flow charts and how the radio resources are kept will make shorter delays for GPRS or EGPRS. In principle, if TBFs have to be established, GPRS provides shorter delays for STS and subsequent STS reflects it. However, the fact of comparing different systems makes it difficult to correlate the values. For VDT, the transmission rate of EGPRS should be translated into a shorter delay than GPRS. Figure 6.23 does not clearly show this behavior, but for the case of a trial using first EGPRS mobiles and afterward GPRS mobiles with the same PoC system, EGPRS provides a reduction in the delay of around 200 ms. From Figure 6.23, iDEN shows a much better performance than (E)GPRS when comparing it with its average value. In the case of STS, the difference is not so big if only the best solutions are considered although the difference in VDT is still high even for the best (E)GPRS values: 1.5 seconds for first STS for (E)GPRS against 0.95 for iDEN; 0.9 seconds for (E)GPRS vs 0.49 for iDEN for subsequent STS; 2.6 seconds for (E)GPRS against 0.32 for iDEN for first VDT; and 1.4 seconds for (E)GPRS vs 0.3 for iDEN for subsequent VDT.

Apart from the matureness of the PoC service in iDEN networks, there is also another important reason for this performance difference of PoC service between iDEN and

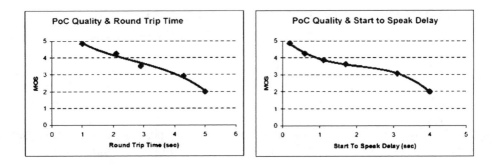

Figure 6.24 PoC MOS quality vs latency

Table 6.11 MOS Quality criteria for PoC users

MOS quality value	Description
5	Delay not noticeable
4	Noticeable but not annoying
3	Slightly annoying
2	Annoying
1	Very annoying

(E)GPRS. iDEN is a proprietary technology optimized for PoC like services, whereas (E)GPRS is not optimized or designed for real time nature and bursty traffic like PoC one. Therefore, most likely, PoC performance in (E)GPRS networks will not catch up the performance obtained over iDEN networks. However, as (E)GPRS networks get enough maturity and some features are included to enhance the PoC service end user performance (extended UL, NACC, streaming QoS . . .) We can expect an important boost on PoC performance over (E)GPRS which could approximate to the performance over iDEN networks.

Figure 6.24 and Table 6.11 show some reference values of what the subjective quality of PoC service (MOS) would be, for different latency values. According to this reference, we can verify that the quality obtained for iDEN is very closed to 5 in most of the cases, which would mean excellent from the delay point of view, while for EGPRS it will be close to a MOS of 4, which is still good quality [12].

6.5 Performance Analysis Example

To close this chapter we have included a performance analysis of the impact that QoS on EGPRS network can have from the point of view of increasing the capacity and providing better performance to real-time users. However, this study is not based on active measurements, but on simulations, as the required network control and testing capabilities were not available. However, the results showed in this section may provide some good hints of the potential gains that could be obtained.

6.5.1 Service Differentiation Impact on Capacity and Performance

The introduction of new and more demanding data services in the wireless networks has entailed to the operators the necessity to put in place algorithms and methods to optimize the efficiency of the limited capacity of their networks. This limited capacity had been evenly distributed among the data users till the appearance of QoS mechanisms, which provide service and/or user differentiation enabling different schemes to share the network capacity.

This section tries to describe how network capacity and performance improvements can be obtained by implementing QoS mechanisms. It will also point out some possible QoS algorithms to be applied in the network as well as some performance improvement figures. For more information on QoS concepts and architecture, please refer to Chapter 4.

QoS prioritization algorithms are used to allow a more intelligent assignment of radio resources. With the term 'QoS differentiation' we refer to methods where treatment of packets in the network is potentially unequal in priority. Some decision criteria are used to determine a priority value for each packet, which is then treated accordingly. Each packet flow belongs to a treatment class. Within one treatment class all packet flows have an equal priority. The objective for differentiated treatment is to achieve better performance according to some metric. Ideally this metric or satisfaction criterion should be closely related to end-user experienced quality of the used service.

There are many decision criteria that can be used to obtain differentiation and therefore, many algorithms can be considered. For example:

- A wise approach would be to differentiate packets in the network according to the type of service. This makes sense since, as stated above, there are several different services offered in the network, and each of them are affected by different circumstances; some services are more critical to delay (conversational or interactive services), other to packet losses (background services), etc. Services tolerating only small delays are assigned a high priority by the prioritization queuing method. The aim of the approach is to be able to carry a higher load with the same network capacity and the same quality of end-user experience (QoE).
- A second approach could be dividing the users into a number of treatment classes according to their subscription type. The motivation is that more expensive subscription entitles to higher priority and thus better service level.
- Another approach to assign the priority value to packets is by selecting the one which best suites to the current radio conditions. This approach leads to increased spectral efficiency, meaning that higher number of user bits are transmitted with given bandwidth and offered load.
- Prioritization of incorrectly received blocks that need to be retransmitted could be another approach. The aim in prioritization of retransmissions is to minimize the end-user experienced delay.

6.5.1.1 Simulation Results

In this section, an example of the first one of the approaches listed above is detailed. By means of dynamic simulations, an algorithm based on service differentiation was tested in a GPRS network [10]. The used prioritization algorithm was a weighted round robin (WRR),

Table 6.12 Service profiles

Service	Service profile
Streaming	Constant bit rate audio streaming. Encoding rate of 20 kbps. A packet with a fixed size of 500 bytes is sent every 200 ms. Application buffer of 8 seconds. The clip length is uniformly distributed between 1 and 5 minutes.
Web browsing	Three fixed-size pages (equal probability, 1/3). The page sizes are 12656, 55353 and 124516 Bytes. Number of pages downloaded per session follows a log- normal distribution with mean value of 4, standard deviation of 5, and maximum value of 10. Reading time between pages is exponentially distributed with mean value of 10 seconds.
MMS	It is considered to be of fixed size of 30 kB.

where there are specific weight parameters for each treatment class. The scheduler grants each user a number of transmission turns per connection which is inversely proportional to the user treatment class weight: i.e. the lower the weight, the more the transmission turns the user receives.

The selected services were, conversational (streaming), interactive (Web browsing) and background (Multimedia Messaging Service, MMS), as described in Table 6.12.

Thus, for every user in each service there is an end-user satisfaction criterion. This satisfaction criterion is required to be fulfilled for at least 90% of the users individually within each service class. The reason behind is that it is assumed that no service can be offered on long term to customers if the quality of that particular service is unacceptable. This is the case even if all the other services would have 100% satisfaction rate.

The satisfaction criterion for streaming service users is that the 8 second application buffer is never allowed to run empty. A user whose streaming client do not suffer any re-buffering is considered satisfied, otherwise it is not.

Satisfaction criterion for Web browsing users is based on maximum downloading time. Each of the three Web pages has their own maximum downloading time, depending on the page size. If this time is exceeded, then that page is considered too slow.

Finally, MMS users are always considered satisfied as long as they send and receive their MMS successfully (Table 6.13).

Also, the traffic mix described in Table 6.14 was chosen for the simulations.

By running simulation campaigns, optimum scheduler weight settings for each treatment class can be obtained. Then, capacity and performance improvements in the system can be quantified. Also, the effect of different traffic mix can be studied. The observed optimum weights are presented in Table 6.15.

Table 6.13 Service satisfaction criteria

Service	Service profile
Streaming	No re-bufferings
Web browsing	Maximum delay on air interface (depending on page size)
MMS	Always satisfied as long as it gets the message

Table 6.14 Traffic mix

Traffic mix (%)	Streaming	Browsing	MMS
Portion of call	1.5	5.4	93.1
Portion of bits	17–18	28–29	54–55

Table 6.15 Optimum scheduler weight settings

Streaming	Browsing	MMS
1 (highest)	4	12 (lowest)

From the results, it was concluded that by using service differentiation, the network was able to carry more than twice the load compared to identical network configuration but with no QoS differentiation. Even sub-optimal weight values produced results with similar magnitude. Traffic mix seems to affect optimal queue weights, although rather significant changes are needed to observe this. When the portion of highest priority class traffic grows, the queue weights start to have smaller effect over the performance of the system. This is normal, since when there is high traffic with the highest priority class, the performance is not limited by the priority settings, but for the number of users within that highest priority traffic class.

In Figure 6.25, the percentage of gain in terms of offered load that service differentiation can produce when comparing with the same network configuration, but not considering QoS, can be observed.

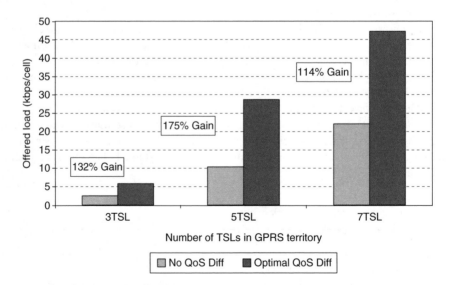

Figure 6.25 Maximum load satisfying streaming and interactive QoS criteria

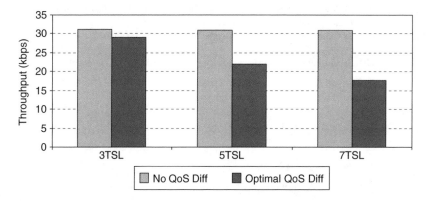

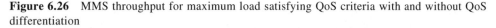

Figure 6.26 MMS throughput for maximum load satisfying QoS criteria with and without QoS differentiation

However, this improvement has a cost, and in this case the fact of introducing QoS differentiation with the optimal weight configuration implies that MMS performance is reduced. The reason is that MMS traffic is assigned with the lowest priority, whereas in the case of no QoS differentiation it has exactly the same priority as the rest of users. Figure 6.26 shows MMS traffic bit rate for the load points shown in Figure 6.25, for both QoS differentiation and no differentiation cases.

References

[1] 'Network Monitoring Fundamentals and Standards', Edmund Wong, ywong@cis.ohio-state.edu.

[2] http://www.optimi.com.

[3] http://www.apple.com/quicktime.

[4] E. Casilari, A. Reyes, A. Díaz and F. Sandoval, 'Modelado de Tráfico Telemático', *Mundo Electrónico*, No. 306, February 2000, pp. 48–52.

[5] E. Casilari, A. Reyes, F. J. González, A. Díaz and F. Sandoval, 'Characterisation of Web Traffic', in Internet Performance Symposium, San Antonio, November 2001.

[6] J. Aracil, 'Características del Tráfico en la Internet e Implicaciones para el Análisis y Dimensionamiento de Redes de Ordenadores', *Novática*, No. 124, November/December 1996, pp. 18–26.

[7] J. Beran, R. Sherman, M. S. Taqqu and W. Willinger, 'Long-Range Dependence in Variable-Bit-Rate Video Traffic', *IEEE Transactions on Communications*, Vol. 43, No. 2/3/4, April 1995, pp. 1566–1579.

[8] W. E. Leland, M. S. Taqqu, W. Willinger and D. V. Wilson, 'On The Self-Similar Nature of Ethernet Traffic', Extended Version, *IEEE/ACM Transactions on Networking Communications*, Vol. 2, No. 1, Febrero 1994, pp. 1–15.

[9] S. Hierrezuelo, 'Estudio de la Calidad de Servicio ofrecida a Servicios de Datos en Redes Celulares', Proyecto Fin de Carrera, ETSI Telecomunicación, Universidad de Málaga, 2004.

[10] A. Kuurne, D. Fernández, R. Sánchez, 'On Service Based Prioritization in (E)GPRS Radio Interface', VTC Fall 2004.

[11] T. Halonen, J. Romero and J. Melero, 'GSM, GPRS and EDGE Performance', Ed. Wiley.

[12] R. Cuny, 'End-To-End Performance Analysis of Push To Talk Over Cellular (POC) in WCDMA', IASTED International Conference. September 1–3, 2004.

[13] www.ethereal.com.

[14] R. Ludwig and R. H. Katz, 'The Eifel Algorithm: Making TCP Robust Against Spurious Retransmissions', *ACM Computer Communication Review*, Vol. 30, No. 1, January 2000.

[15] Clint Smith and Curt Gervelis, 'Wireless Network Performance Handbook', McGraw-Hill, Hardcover, Published May 2003.

[16] 3GPP TSG 44.060, 'Radio Link Control/Medium Access Control (RLC/MAC) protocol'.

7

Customer Experience Management

Brian Carroll

Customer Experience Management (CEM) is an innovative approach to Service Management developed in conjunction with leading wireless operators. It focuses on directly measuring the customer experience in real time and defining Customer Experience Indicators (CEIs) that can be used to actively manage the Quality of Service (QoS) being offered to subscribers. In this chapter, we review the need for service management and compare some of the different approaches to this problem: *Classic Service Management*, *Active Service Management* and *Customer Experience Management*.

By analysing the categories of Key Quality Indicators (KQIs) that each approach provides, we see that there are significant differences in the degree to which actual, auditable customer experience is captured by these methods. When these KQIs are used to define external customer Service Level Agreements (SLAs), these differences can result in service-level violations that cannot be related to actual customer experience.

7.1 Overview of Customer Experience Management

The continued growth of mobile industry revenues depends to a large degree on the adoption and success of mobile services (Figure 7.1). In order for an operator's offering to be truly successful, the shift from managing networks to managing customers needs to happen. Mobile operators must actively monitor and manage the experience their customers have while using their services. A high-quality experience will influence positively on the customer's view of the operator brand and enhance its value.

Unfortunately, today's networks do not allow the definitive QoS experienced by mobile customers to be monitored and therefore managed. Network operators have identified this as being one of the key challenges facing the industry today. This challenge can be broken down into two main areas.

End-to-End Quality of Service over Cellular Networks: Data Services Performance and Optimization in 2G/3G
Edited by G. Gómez and R. Sánchez © 2005 John Wiley & Sons, Ltd

Complexity is Continuously Increasing

- Customers
- Customer groups
- Service types
- Cell configuration
- Handset types
- Billing rules
- Provisioning

2004 +
Mobile Multimedia

| Training |
| Surveillance, Security |
| High-Speed Web Browsing |

2000 +
Data Services

| Advanced Corporate Apps |

Video Browsing	Email with attachments
Corporate Applications	RT Stream: News Sport, TV, Advert
Email Text	News, Sport, Weather, Traffic
News, Sport, Weather, Traffic	Video Ringtones, Multiplayer Games
Ringtones, Java Games	Ringtones, Java Games

Late 1990's
Textual

| | Messaging, SMS, Photos | Messaging, Video, Chat |

Mid 1990's
Telephony

| | Messaging, SMS | GPRS | Videophone |
| Voice | Voice | Voice | Voice (GSM, UMTS) |

Data kate

Complexity & Time

Figure 7.1   Customer Experience Management

1. No real-time view of the actual quality delivered to the customer.
2. No mechanism to target network operation activities towards the most valuable customers.

CEM addresses these challenges head-on. It collects the vital statistics with the granularity needed to efficiently manage these services and ensure a high-quality experience. CEM turns these statistics into customer experience knowledge as it allows the operator to build an understanding of the customer's definitive experience based on a number of CEIs. It provides an understanding of how the customer's experience affects behaviour and provides the input needed to optimize the network performance where it matters most to high ARPU customers.

The features of CEM will benefit a number of functions in the operator organization.

- *Network quality department*: Enables accurate monitoring of the network and service quality that was actually delivered to a customer or group of customers and provides the information needed to bridge between the perceived quality and the quality actually delivered.

- *Operations*: Provides information on where to fine-tune the network for an enhanced customer experience and enables the focus of the department's efforts to be set according to the corporate focus of the operator.
- *Marketing & Sales*: Adds context to existing network statistics and customer satisfaction surveys correlating it with usage statistics per customer, service and location. It provides support to the process of building a brand experience and also provides account managers with detailed reports on the actual level of service provided to the accounts.
- *Customer Care*: Allows Customer Care to be proactive in their interaction with customers rather than re-active in waiting for a customer complaint and enables the build-up of customer profiles.

7.1.1 The Challenge

The mobile services market has been hyped to the extreme, and CEM is the only viable way to regain control of the customer's experience and manage it moving forward. Traditionally, mobile operators to a large degree have managed their operations based on a network or technology view of the network. However, as customer growth levels off and prevention of churn becomes critical, many operators are realizing that the network view alone is not sufficient for their business.

In order to drive widespread adoption, it is important to greatly enhance the quality of the customer's experience of these new services. Today's networks do not allow the quality of the customer experience to be monitored and therefore managed, and network operators have identified this as being one of the key challenges facing the industry today. The challenge can be broken down further and be described as being:

- *To measure QoS as experienced by customer in real time*

  - Gap in the statistics available and necessary to measure customer-based QoS
  - Lack of tools supporting this approach

- *To guarantee customers a satisfactory level of QoS*

  - Lack of tangible input for successful SLA management

- *To manage customers' QoS expectations*

  - Lack of sufficient measuring methods and management tools
  - No mechanism to target operations towards a specific set of users

- *To succeed in launching services*

  - Lack of methods to manage the initial subscribers
  - Lack of tools to ensure that customers' demands are met and that the mobile services generates revenues.

7.1.2 The Solution

The CEM technology offers a solution to this problem. Features and benefits delivered by this technology are as follows.

- *Measure QoS as experienced by customer*
  - Provides customer-centric, real-time measurements
  - Allows QoS per service to be associated to customer and location

- *Guarantee customers a satisfactory QoS level*
  - Provides crucial input to the SLA process
  - Enables the establishment of customer profiles
  - Provides Customer Care and Marketing & Sales with information of how the customer perceives the QoS

- *Manage customers' QoS expectations*
  - Provides the management tool needed to offer customers a satisfactory level of QoS

- *Succeed in the launch of mobile services*
  - Exceed customer expectations
  - Enhance customer satisfaction and loyalty
  - Turn customers into advocates for the operator's offering
  - Increase ARPU.

By adopting a CEM approach, the operator can gain a detailed and accurate real-time view of what the customer actually experience when using a particular mobile service. This information is a prerequisite for any process aiming to improve the customer's view of the mobile service and the quality delivered.

These statistics include, for example: Was customer Mr Smith able to use the services on your network when he wanted to? Did the service provide the level of quality that he expected? Was he able to complete his business and exit the service successfully?

CEM turns these statistics into knowledge as it allows the operator to build an understanding of not only what the customer experienced but also when, why and with what equipment/handset. It also provides the input needed to optimize the network performance where it matters most to the high ARPU customers. In addition, it allows the operator to understand how the customer's experience affects his or her behaviour. This kind of information cannot be readily extracted from the statistics available in today's networks.

7.1.3 Driving Mobile Revenue

CEM provides an operator with the information and tools needed to focus its network management on the most important factor of all, i.e. the actual customer experience. It allows an operator to quickly detect and fix any problems in the network before they impact negatively on the user experience. It also provides a benchmark to drive continued quality improvement.

It gives an understanding of how a mobile service is behaving in the eyes of a customer, and the improvements that needs to be made to increase the quality of the experience.

By directly managing the customer experience, operators now have the capability to:

- Turn existing customers into advocates for mobile data technology in general and the operator's offering in particular
- Acquire new high-value subscribers to its mobile network

- Retain existing customers and encourage the use of additional services
- Increase ARPU.

7.1.4 Maximizing Operational Efficiency

Network operations needs to be directed towards the area of most 'pain' for the operator and to the maximum benefit to the overall business.

CEM enables problems in particular areas of the network to be pinpointed. Threshold-based alarms warn the operator of impending problems whilst historical archives of performance data allow rapid response to customer calls and fast closure of trouble tickets.

By guiding optimization engineers in their efforts to fine-tune the network, CEM also mitigates against problems occurring in the future, offering further efficiency improvements.

7.1.5 Enhancing Customer Care

As today's networks do not allow network operators to view the network and service performance as the customer sees it, it is not possible to take a fully proactive approach to Customer Care.

Even if the operator is aware of a problem in the network, which may have had a negative impact on its customers, it is not until a user has logged a complaint about his or her poor user experience that the operator is made aware of the extent of the problem for this particular customer.

This leads to a re-active Customer Care process, a problem-oriented communication with customers and an unnecessarily high call volume to the Customer Care unit, with the subsequent high operational costs (as shown in Figure 7.2).

The ability to offer fine-grained customer care and accountability for service levels are extremely valuable tools in retaining existing customers and acquiring new customers. Using CEM, customer-affecting performance problems can be seen in real time, allowing Customer Care to be proactive in their interaction with customers, rather than re-active in waiting for a customer complaint. In doing so an operator will reinforce its strong commitment to

Figure 7.2   Re-active Customer Care

Figure 7.3 Proactive Customer Care enabled by Customer Experience Management

customer care and build customer satisfaction and loyalty. This will also reduce the volume of incoming calls to the Customer Care unit, leading to increased customer satisfaction and lower operational cost (Figure 7.3).

CEM focuses on helping operators meet their business goals by enabling them to monitor the service offered from the customer perspective, not just the network perspective. This will facilitate improved product offerings to customers with per customer QoS guarantees. By enhancing the customer experience, the achievement of higher retention rates and differentiated product offerings is now feasible.

7.1.6 Measuring and Building Customer Satisfaction

In today's highly competitive telecommunications landscape, increasing and maintaining a high level of customer satisfaction is of paramount importance to a mobile operator. A customer's overall satisfaction with the service offering is influenced by a number of factors.

One main contributor to the satisfaction is the actual experience of using the mobile service, for example the speed of connecting to a data service. Today, the view of the customer's satisfaction is mainly made up of a combination of the result of customer surveys and various network measurements. The output does not provide the granularity nor the exactness needed to build an accurate view, and it does not provide the high-quality input needed by the operator's commercial and technical management.

CEM adds context to this data as it enables the satisfaction rates expressed by customers to be correlated with detailed usage statistics per customer, service and location. By extracting knowledge from the statistics, it provides the commercial and technical management with explanations to the reasons for the achieved customer satisfaction rates and information on where to focus the resources to improve satisfaction.

CEM allows Customer Care to adapt a proactive approach, as described earlier, and contact the high-value customer who experienced poor quality during a recent mobile service usage to inform them about what actions are being taken to compensate the customer for the poor experience and what actions are being taken to avoid this happening again. This level of customer focus and rapid response will prove to the customer that he is indeed the operator's most valuable asset, which subsequently will build customer loyalty.

7.1.7 Building a Brand Experience

Managing the experience a customer has with a specific mobile brand has become increasingly important in keeping customers in the continued consolidation in the market.

The perception of the quality of an operator's services is a main contributor to this brand experience. CEM provides an operator with an invaluable supporting tool when building a brand experience. This includes support to define what experience the customer will have with the mobile service offering, and support to ensure that this experience is delivered to the customer.

CEM allows the definitive customer experience to be measured in real time and with high granularity and precision. This functionality will take the uncertainty out of the customer experience measuring process and bridge the information gap between the satisfaction rates expressed by customers and the high-level network statistics.

7.1.8 Improving Network Quality

One important benefit of enhanced visibility of the customer's experience of the mobile network is that it will steer the network optimization tasks to the areas that will improve the overall experience. It allows for rapid identification of problem areas in the network and expedites problem resolution.

This ability to continuously optimize the network and decrease mean time to problem resolution, leads to a network with a high QoS and ultimately to satisfied and loyal customers and greater product differentiation. In early GSM networks, a positive side-effect of increased confidence in the quality of a mobile service being offered was the increased usage of that service.

In adopting CEM, an operator can secure the success of their mobile services and revolutionize the way these services are managed and perceived by their customers.

7.2 CEM and Service Management

7.2.1 The Need for Service Management

Service Management is not a new concept in the Operation Support System (OSS) world and indeed the category has been recognized for several decades by standard bodies such as the ITU-T [1] and TMF [2] as well as by the wireless and fixed telecommunications industry in general. However, it is only in the last few years that operators have started to look again at

Service Management with a renewed urgency. The reasons for this lie in several fundamental business trends that are starting to force a new assessment of how best to manage the services offered to customers.

Perhaps the most significant factor is the emergence of wireless data technology such as GPRS which provides a reasonably fast and 'always on' bearer over which a multitude of wireless data services can be offered from Internet access and email through to corporate VPN and consumer-focused location-enabled games. What is significant is not the bearer itself, nor specifically the services, but the customer experience gap between the bearer and the services, as illustrated in Figure 7.4.

The QoS experienced by the user of, say, a multimedia football final score service is dominated by the timeliness of delivery, the quality of the content and other factors not always simply related to the performance of the underlying bearer (see also Chapter 4). In truth, this gap has always been present even for basic voice services but in this case it is so much smaller that the difference between managing the quality of the bearer and managing the QoS has not been significant. This has now clearly changed. While assuring the basic availability and retainability of the GPRS network is still essential, management systems that focus on the network provide extremely limited visibility of the actual services that are really being experienced by their customers. It is this gap that has led to a new focus on Service Management for GPRS and 3G.

Financial pressures in the industry are the second key factor in the new importance of Service Management. Senior Executives of Wireless Operators are under enormous pressure to demonstrate that their stock justifies its continued growth rating and that they are not simply becoming high-technology utility companies. This fuels a need to develop more services that can fill the revenue gap caused by falling or flat ARPU and continuing high levels of

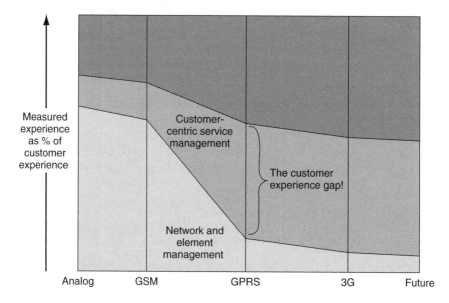

Figure 7.4 The growing customer experience gap

churn. Operators have to become focused on services and, critically, become customer facing rather than network facing.

Customers buy services, not networks or technologies, and therefore operators now fundamentally must shift to managing the product they actually sell, i.e. the service, rather than the technology that delivers it, i.e. the network. Service Management solutions potentially fulfil that need. Such factors are principally drivers of Service Management for GPRS and 3G.

However, almost as strong factors are making operators look at Service Management, and especially CEM, as an important tool for existing GSM networks. GSM technology for voice is essentially mature and well understood, and network and element management systems have been in place for many years that assure the availability of the network. However, more and more operators are finding that they have reached a 'glass ceiling' in terms of how much better they can make the network using existing tools that only provide visibility at an element level, i.e. per cell performance data. What they need is to directly measure what individual customers are experiencing, where in the network they have problems and what detailed faults and performance issues occurred. This is the only way to break through consistently less than 1% dropped call rates or close to 100% availability levels. But does this focus on the last few percent make financial sense? The resounding success of GSM means that it does.

Even in medium-sized networks that final percentage point represents possibly hundreds of thousands of voice calls or SMS and very sizeable revenue that is potentially being lost to the operator. Perhaps more significantly, premium subscribers are probably more likely to be experiencing these problems if only because of their substantially higher usage of the network. In an environment of intense competition, measuring the customer experience, particularly for high-value customers, is a vital part of customer retention and increasingly a strong differentiator when competing for corporate customers.

7.2.2 The Service Management Landscape

Three distinct but complementary approaches have emerged to provide Service Management Systems to wireless operators, what we shall call 'Classic' or Synthesized Service Management, Active Service Management and CEM. To a large extent the first two are evolutions of existing network management approaches, namely network performance and fault management, and network test and measurement. The latter is a more radical approach to meeting the needs of the new customer-centric OSS.

Of course practical deployments often mix elements of all three of these approaches usually with the emphasis on a particular one and we shall consider such an approach later in this chapter. But it is valuable to consider each approach individually to fully understand their strengths and weaknesses and to gain some insight how they can be combined together in a complementary way.

7.2.2.1 Classic or Synthesized Service Management

The Classic Service Management approach is best represented by the work of the TMF with regard to Wireless Service Measurement [3]. In this approach, product managers are asked to define customer-focused Key Quality Indicators (KQIs) that are specifically relevant to their product or service.

These KQIs are customer focused in the sense that they should cover the broad customer lifecycle from product acquisition at point of sale, through provisioning to actual usage of the product in the network; and are phrased with reference to a model of the customer experience of the service.

Once such KQIs are defined, existing network KPIs such as those derived from fault, performance and billing data are analysed and algorithms are defined to synthesize the KQI from measured KPI. Service Management is thus layered on top of Network Management as classically envisaged by the ITU-T in the TMN standards (Figure 7.5).

Such a Service Management solution demands a high degree of flexibility in the OSS as data must be combined together in a flexible and coherent way, accommodating the fact that it may be event or period based, the periods may differ, and the reporting elements may be different and so on. In general, weightings have to be applied to each KPI, based initially on the analyst's best estimates, to provide a KQI that represents the underlying importance of each constituent KPI. Once the service is launched, these KQIs can be tested through customer research and the algorithms refined – a process that is clearly not without significant cost.

A lot of early work on Service Management was focused on fixed networks – in particular SDH networks. However, such networks have two particular characteristics that make them much simpler to manage from a service perspective compared to a wireless network.

By its very nature, the end points of a fixed network are fixed and therefore it is possible to get an end-to-end view by installing monitoring equipment at the customer Service

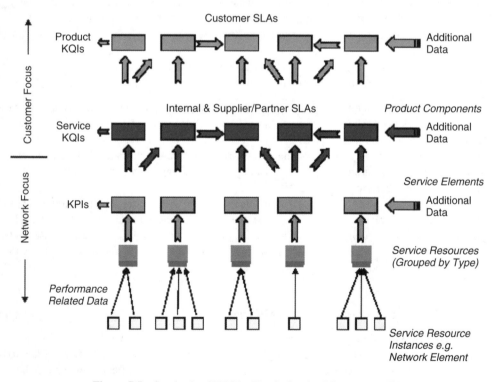

Figure 7.5 Synthesis of KQI in Classic Service Management [2]

Access Point (SAP). This is not the case in a wireless network where the customer SAP, the mobile terminal, is highly mobile and, at least with today's technology, is not able to support a complex monitoring application. In addition, the path taken by a service bearer through the network is less complex and it is therefore easier to relate the service performance to underlying network performance indicators. In an SDH network, the network itself directly measures the end-to-end performance of virtual circuits using O&M overhead in the data packets making Service Management by this approach even more viable. By contrast, the bearer path in a wireless network is almost totally determined by the location of the subscriber (defining the path through the BSS/RAN and SGSN/VMSC) and the service selected (defining the GGSN and service platform such as WAP gateway, MMSC and GMLC). Matching specific network element performance to a specific service instance would therefore require very dynamic information about routing and therefore is often only attempted at a much more gross level.

Synthesized Service Management in a wireless network is thus an extremely difficult and challenging task but one with potentially high rewards.

7.2.2.2 Active Service Management

Active Service Management is a natural evolution of existing test and measurement systems that are used by network operators during service rollout and for service test and validation. Such systems measure the performance of services by actively invoking the service and measuring KQIs directly based on the responses received from the network. Service tests can be made highly configurable and fully automated to allow regular and consistent measurement of service performance through the network (see section 6.3 for further details on the trial methodology generally followed when using this type of solutions).

Because these solutions are working directly on the application layer and are often connected to the network at the same access point as real subscribers, for example via a terminal over the air interface to the BSS/RAN, they provide a very direct means of measuring the performance of a service as it might be experienced by a subscriber.

However, this same advantage is a two-edged sword. While active testing, say HTTP to a specific service URL over IP across a GPRS connection, can provide genuine visibility of the end-to-end performance and availability of a wireless Internet application, the very nature of the layering of the protocols makes it hard to understand where and why problems may have occurred. That is why active test solutions are sometimes complemented by other technologies, such as IP sniffing at different level and/or passive probes at different points in the network to support fault diagnosis, or simply used as way to automatically monitor, report and trigger real-time alarms for SLAs.

Advantages of this type of solutions are the multi-use and multi-technology flexibility, allowing to provide systematic service performance capability that measures all services independently on the network and terminal technology. In addition, it provides increased network visibility and accuracy in the performance results on specific areas of interest for the network operator (e.g. busy traffic location).

Additionally, active testing allows going down into the impact of radio quality by correlating application level measurements with tools capable of measuring signal levels and signalling information from the network during the whole test.

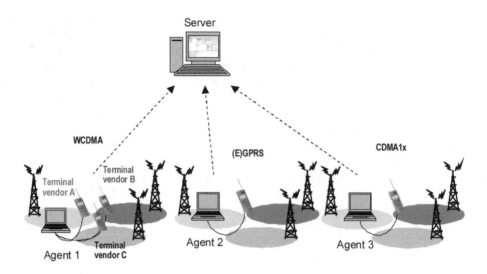

Figure 7.6 Active Service Management approach

However, scalability is a significant challenge since it is not possible to actively test each service through every possible service access path, i.e. every cell. Consequently, true end-to-end deployments of active test solutions are limited to main areas of interests (key cells) and sometimes deployed in the network, for example at the A-interface for voice call management or in the IP core for data services, to provide a scalable solution.

An example of Active Service Management approach is illustrated in Figure 7.6.

7.2.2.3 Customer Experience Management

CEM is a new category of Service Management, which attempts to directly measure the customers' experience of the quality and performance of services in real time. By customer experience we mean the throughput, delay or error conditions that were actually experienced by a single subscriber – factors that are likely to directly impact their perception on the quality of service. Customer Experience Management shares some characteristics of the synthesized approach in that it is passive, i.e. does not actively invoke services but passively monitors real subscriber usage; and it is pervasive, i.e. it monitors all, or most, customers to some degree everywhere and all the time.

However, unlike the synthesized approach it directly measures real-time performance on a per customer or per customer group basis rather than attempting to infer it from network-based measurements. Similar to the active test approach, near end-to-end measurements can be made at the application layer and also retaining the ability to drill down into the underlying protocols, for example BSSGP, GTP, to support accurate fault diagnosis. For example, CEM uses probing technology to monitor in real time the Gb, Gn and A interfaces and collect the actual customer experience of GPRS and GSM services (as shown in Figure 7.7). Monitoring around the SGSN gives a good compromise between

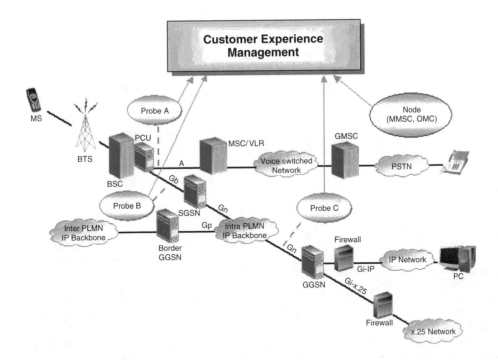

Figure 7.7   Customer Experience Management for GPRS/GSM

seeing a full end-to-end view of performance and scaling to a pervasive solution for the entire network.

However, the CEM vision is that ultimately the handset itself will evolve to become the ideal point for passive monitoring of the customer experience. The handset is as close to the customer as possible and is able to directly measure accessibility indicators, such as holes in RF coverage or accessibility of GPRS, which simply cannot be seen by the network itself. While some agent-based solutions exist today, they are generally limited to the user layer and so are unable to link problems here with the underlying mobility management, session management and RF measurements. Trials of handset-based customer data collection technology in an existing GSM network have already been conducted, and, as the application environment in the handsets evolve, will continue to develop solutions in this area.

A key feature of passive monitoring, unlike an active test solution, is that it is able to see down through all the protocol layers thus providing visibility of not only the user layer data such as the HTTP or WML application requests, but also the BSSGP signalling in the control plane or PDP context information carried over GTP. Consequently, when a KQI does indicate a service problem, additional information can be provided in the form of low-level signalling error codes and related network data such as current cell or link information. This allows the operator to drill down using existing network performance systems to understand what is causing the service problem, combining the direct measurement advantage of the active approach with the fault diagnosis capability of the synthesized approach.

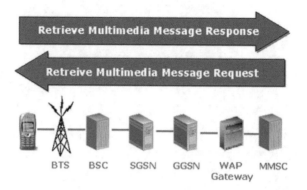

Figure 7.8   Example MMS retrieval KQI

7.2.3 Categorizing KQIs by Customer Experience

A key capability of all Service Management solutions is the creation of KQIs, which are used as the basis for both the management of services and customer experience and as the building blocks for SLAs.

However, while each solution may have similarly sounding KQIs, the differences in the approaches lead to a fundamental difference in one aspect of the KQI – the degree to which they represent actual customer experience.

To see this, let us consider one example KQI for a Multimedia Messaging Service (MMS). Suppose our analysis of the customer experience suggests that an important indicator is the ability of a subscriber to retrieve a multimedia message using their MMS phone that has been sent to them by another subscriber. We therefore define a KQI as the MMS Retrieval Service Success Rate being the ratio of the number of successful retrievals to the number of attempts initiated from the MMS terminals (not simply received at the Multimedia Messaging Service Centre, MMSC). This scenario is depicted in Figure 7.8.

This is a complex end-to-end process involving the BSS radio and GPRS core networks, the WAP and IP infrastructure and the MMSC. Let us consider how and what each Service Management solution might measure.

7.2.3.1 Active Service Management: Implied Customer Experience

Suppose we have an Active Service Management solution that periodically actively attempts to send and retrieve a multimedia message from the MMSC. Where the system is deployed, either across the radio interface or in the core IP network, will determine how much of the end-to-end performance will be included in the KQI. If the system is deployed in the IP network, then problems in the radio network or GPRS core network will be invisible to the system and will not be reflected in the KQI. But in both cases, the system measures only what it experiences and not what subscribers experience – we call this *Implied Customer Experience*.

The experience of the test system is used to infer that subscribers are probably also having this problem. For gross failures the implication may be acceptable, but for a service that is not heavily used or where the poor availability is only intermittent, the system may be recording a KQI that is not in any sensible way reflecting what subscribers are actually experiencing. The growing number of services and service options also brings the risk that the system is not testing sufficient options. In our example, subscribers who access the MMSC using an URL that initiates a 'retrieve and delete' function as opposed to a simple 'retrieve' might be seeing failures but this will not be recorded by an active test solution unless it covers that specific test case.

7.2.3.2 Synthesized Service Management: Non-Specific Customer Experience

Depending on the availability of data, the Synthesized Service Management System may attempt to define the KQI using performance statistics available from the MMSC, the GPRS network and the BSS.

These raw KPIs would be weighted, probably with the per element MMSC retrieval success rate dominating, to define the required KQI. While in a sense providing an end-to-end view, at least from the BTS onwards, the problems of mobility means that without substantial complexity a real end-to-end view is difficult to accommodate. Several dozen cells or even an SGSN in rural Wales might fail and so impact the KQI and yet there is no MMS usage in the area. Conversely, a single cell covering a wireless village conference might fail and barely be reflected at all in the KQI, yet severely impact many customers' experience of the service. Although billing or other usage data can be incorporated to help in such an analysis, it rarely has the granularity needed or the real-time availability.

The same 'bluntness' applies to whether the KQI reflects the customer experience or not. Clearly, actual peg counts of retrieval failures at the MMSC or PDP context failures to an MMS APN do indicate actual customer experience of MMS retrieval failure. However, there is no information on which customers experienced the problems (were they prepaid subscribers or high-value business subscribers?) and no opportunity to drill down on a per subscriber basis and understand the actual error codes they were receiving when attempting to access the service – we call this non-specific customer experience.

7.2.3.3 Customer Experience Management: Specific Customer Experience

Depending on where data collectors are deployed in the network, the CEM system may see a partial end-to-end view of performance. For example, monitoring at the Gb interface provides visibility of the GPRS network and IP network and a limited view of the radio network from error codes passed upstream on the signalling plane. In our example, its KQI will be directly derived from aggregating actual failures to the MMS APN and attempts to access the MMSC for retrieval that result in failures. Each failure represents a specific customer experience and the KQI is quantitatively derived from these actual failures with no need for further weighting or tuning – it can be audited directly back to actual messages monitored on the relevant interfaces. This category of KQIs which is directly and auditable related to individual or groups of customers is called *specific customer experience*. To separate specific customer experience KQIs from other KQIs that are not direct, actual measurements of such experience have defined the new term *Customer Experience Indicator* or CEI.

7.2.3.4 Impact on Customer Specific SLA

One of the functions of a Service Management solution is to support SLAs between various 'actors' in the value chain. These SLAs may be internal between, for example, network operations and a service provider organization or they may be external between the service provider and the subscriber. Up until now the use of external customer SLAs in wireless has been rare, except perhaps at a high level for large corporate. But in the current harsh competitive climate where most European markets are close to saturation, competing means taking another operator's customers and making sure you keep your own.

Operators and service providers are finding that external customer SLAs can potentially become potent sales differentiator. But what sort of SLAs can actually be provided by the various Service Management solutions that we have considered?

All Service Management solutions allow the definition of SLAs in terms of KQIs that are being monitored and possibly in terms of other internal SLAs. This means that with regard to customer experience they share the same categorization as the base KQI: implied customer experience, non-specific customer experience and specific customer experience. As external SLAs may be part of a binding contract between the operator and the customer and may have financial consequences if they are violated, it is important to understand exactly the implication of these categories, which are listed in Table 7.1.

Table 7.1   Categorizing KQIs by customer experience

Category	Description
Implied customer experience SLA	If such an SLA is broken, it means 'we believe that if the customer had attempted to use the service at this time they would have suffered service degradation'. There is no way of knowing that they did experience any problems and, indeed for low usage services or intermittent problems, no real subscribers may have actually experienced the problem. For SLAs that are concerned with service integrity and retainability, i.e. that imply usage, it seems overly strict to risk financial penalties on the basis of an implied poor customer experience when it is perfectly possible to measure if the problem was actually experienced. On the other hand, for SLA that are concerned with service accessibility, i.e. can a subscriber access the service in the first place, the implied customer experience may be the best that can be achieved because of the difficulty of measuring whether customers are having problems accessing a service from an end-to-end perspective.
Non-specific customer experience SLA	In this case, if such an SLA is broken, it means 'some customers have probably experienced this service degradation and this customer may have been one of them'. Again, the operator, without other data available, does not know that this customer experienced poor performance and may have to forego penalties to all customers with this product SLA even though only some of them are aware of the problem.

Table 7.1   (*continued*)

Category	Description
Specific customer experience SLA	If a specific customer experience SLA is broken, it means 'this customer (IMSI group) actually experienced this degradation of performance'. SLAs based on CEM, which monitors per subscriber and per subscriber group usage of services, are specific and directly auditable to real customer experience. Only those SLAs for customers who experienced the problem are violated and incur penalties. This sort of SLA works best for integrity and retainability SLA where there is measurable usage. For accessibility SLA, in the absence of a handset-based solution, careful analysis must be undertaken to ensure that sufficient cases of inaccessibility can be detected – this will typically depend on how close to the radio network the end-to-end path can be monitored.

7.2.3.5 Hybrid Customer SLA: The Optimal Approach?

The above analysis is intended to help understand the tradeoffs between the fundamental approaches by examining their impact on customer SLAs. While some operators will implement only a single approach, many operators will have multiple Service Management solutions. We believe there is an opportunity to enhance the overall solution by combining CEM with the other approaches.

Our conclusion from the analysis of these categories is that the combination of Implied Customer Experience SLAs and Specific Customer Experience SLAs provides the best balance between proactive detection of potential service violations and auditable measurement of actual customer experience. In particular, SLAs based on the accessibility of services are probably best derived from implied customer experience, whereas SLAs based on retainability and integrity should be derived from specific customer experience.

However, a CEM system that employed handset-based data collection would be able to directly provide specific customer experience SLA for accessibility indicators without the disadvantages of the active test approach and probably represents the optimum long-term solution.

There are clear opportunities to combine the approaches to produce hybrid customer experience SLAs. For example, a sophisticated active test system might be used to test a payment scenario in an m-commerce application – perhaps the receipt of payment confirmation – and an internal SLA is defined for this use case.

A CEM system might not be able to detect the particular flow through the application logic for this scenario but could perhaps measure the number of subscribers redirected to an error page and an internal SLA be defined accordingly. These two internal SLAs can then be combined with an external SLA that is characterized as part implied and part specific. The SLA means, 'We know this customer is having problems with this m-commerce application and our tests imply it is with payment receipt.' The architecture of such a solution is considered in the following section of this chapter.

7.2.4 Architecture Options for Customer-Centric Service Quality Management (SQM)

In the following discussions on architecture, it will be assumed that an SLA management component is separate from the Service Monitoring component and that it has published external interfaces. Some vendor products may combine SLA and SQM together in a single component, which limits the options for the integration of complementary solutions.

7.2.4.1 Combining CEM with Active Service Management

The following diagram in Figure 7.9 shows how CEM can be integrated with Active Service Management at the SLA level in order to gain the benefits of both approaches in terms of implied and specific customer experience SLAs.

In this architecture, the two systems are combined through the SLA management component, which combines internal SLAs received from either system with an external customer SLA. The interface would typically be some standardized alarm IRP such as ETSI TS 132.111 using CORBA or something similar.

The CEM system monitors actual customer experience through the customer experience data collectors and stores this data in a database aggregated across defined IMSI groups. KQIs are defined in the system based on this data and are monitored for selected IMSI groups, for example large corporate or inbound roamers, against configurable thresholds. Any threshold violations can be directed internally to operations or engineering teams or can be forwarded to the SLA management system as internal SLA violations. Similarly the Active Service Management System runs fully automated tests in the network and can forward SLA

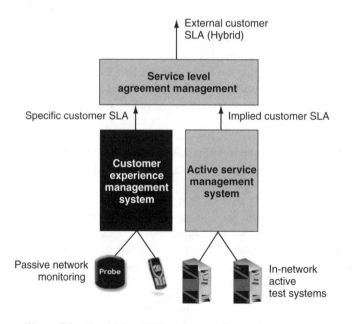

Figure 7.9 Combining CEM and Active Service Management

violations upstream to the SLA management system. These can be combined to produce hybrid customer experience SLAs as discussed earlier.

This architecture has a single point of integration, which makes it more cost effective to deploy. Further integration may be desirable, for example, to allow operators to drill down in the CEM system based on alarms or reports from the active test system. Such integration could initially be cost effectively performed at a user interface level rather than by sharing data.

7.2.4.2 Combining CEM with Synthesized Service Management

Currently, there have been very few implementations of Synthesized Service Management in wireless service providers because of the technical issues outlined earlier, and the potentially high cost of integration with all the underlying network and element management systems. However, where such solutions have been implemented, CEM can still add significant value by adding real customer experience to the defined service or product SLA, and by improving, or indeed replacing, the existing KQIs by the addition of actual measured CEIs.

The following diagram in Figure 7.10 shows how CEM could be integrated with Synthesized Service Management at both the SLA management level and the KQI level to provide more specific customer experience data to the Service Management solution.

Again, we see that the principal benefit of introducing specific customer experience to the external customer SLA can be easily achieved by integration at the SLA management system. In addition, data from the CEM system can be integrated directly into the Service Management System allowing new specific customer KQI to be synthesized. This approach may only be possible if the Service Management solution is able to represent individual subscribers and subscriber groups as managed objects, as opposed to just network elements and products/ services – this being what the CEM KQI directly relate to.

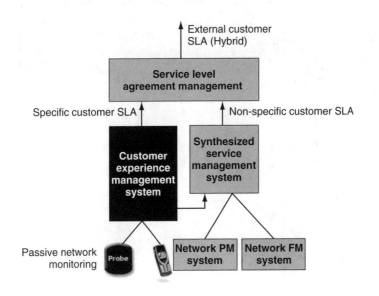

Figure 7.10   CEM and Synthesized Service Management

7.3 Advantages CEM Brings to an Operator

CEM introduces the ability for mobile network operators to monitor and manage the QoS experience of customers' use of their services all the time, in real time. CEM has identified and implemented data sources to capture, aggregate and model this experience for all customers.

CEM provides monitoring of the accessibility, integrity and retainability of the mobile network and its services as experienced by all of the customers all of the time. The experience of all customers is monitored by collecting information directly from the network and providing detailed information about key aspects of the customer experience which is captured by:

- Individual customer
- Group of customers (including prepaid, postpaid, corporate, etc.)
- Group of groups
- Corporate, SME, SOHO
- User equipment
- Services
- Access Point Name
- Location.

Figure 7.11 shows the added values that CEM provides to a network operator.

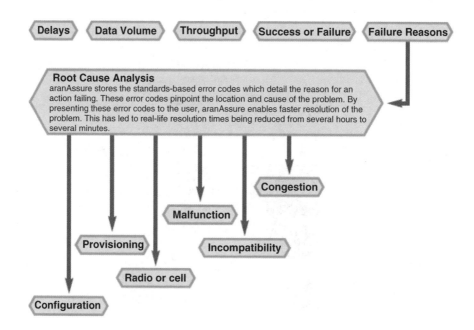

Figure 7.11   Added values that CEM provides to an operator

The following benefits are also provided by CEM.

- Alarm on any/all customers experiencing poor QoS.
- Once unacceptable experience is identified, then provide easy access to key diagnostic information like fault codes and cell information to minimize resolution time.
- Provide scheduled or on-demand reports of customer experience to stakeholders.
- Quickly and efficiently navigate customer experience information for customer groups, services and root causes; drill down to individuals to aid diagnosis.
- Efficiently investigate and diagnose customer complaints.
- Use CEM information to differentiate the service offer in the market through experience-based SLAs.
- Adapt to your need in terms of functionality, cost and evolution over time.
- Design to scale to the standard Tier 1 requirements now and in the future.

7.4 Summary

In this chapter three main approaches to Service Management and how they lead to different categories of KQI and SLA concerning non-specific and specific customer experience have been discussed.

The value of measuring specific customer experience leads to take into consideration how a CEM system can be combined with an Active Service Management system to provide to wireless operators a comprehensive customer-focused approach to Service Management. This solution combines the proactive measurement of accessibility and other radio and user indicators, with specific and auditable measurement of customer experience, to deliver an ideal solution for the management of present and future data services.

References

[1] International Telecommunication Union (ITU), Telecommunication Standardization Sector, http://www.itu.int/ITU-T/.
[2] TeleManagement Forum, www.tmforum.org.
[3] Wireless Service Measurement Handbook, GB923 V3.0, TM Forum.

8

Service Performance Optimization

Gerardo Gómez, Juan Torreblanca and Mattias Wahlqvist

8.1 Introduction

The optimization of the end-to-end service performance is not a simple task since many different protocols, procedures, network elements and interfaces are jointly involved in the services scenario. Important achievements have been made in L1/L2 optimization in wireless networks, like new modulation schemes or error protection techniques. However, the optimization of the service performance requires an end-to-end solution, from physical to application layers.

As explained in previous chapters, some of the protocols and application contents commonly managed in fixed networks and currently reused in mobile network elements are not initially adequate to inherent wireless characteristics (due to long delays, variable bit rates, wireless terminals characteristics, etc.). This issue has led to the development of a set of optimization techniques at different levels oriented to optimize the end-user performance.

This chapter is intended to provide some guidelines on how to optimize the service performance, describe the main techniques available for that purpose and quantify the performance gain from some of them.

The whole end-to-end optimization processes may be applied at different levels and locations in the transmission path of the data. They can be grouped as follows.

- *Network-level optimization methods* – include all processes applied in the network, oriented to optimize network parameters, architecture or dimensioning in order to improve the network Key Performance Indicators (KPIs), such as the Round-Trip Time (RTT) or the throughput.
- *Transport-level optimization methods* – are mainly oriented to TCP optimization, although some research is also focused on WAP- or UDP-based services. The objective is to minimize the impacts that high delays and data losses have on transmission rates and response times over TCP.

- *Compression techniques* – include all mechanisms oriented to reduce the amount of data to be transmitted through the air interface. They can be applied to the data itself or to the protocol headers.

TCP over wireless has been one of the most important research areas during last years. Despite the fact that TCP is a peer-to-peer protocol (from terminal to server or terminal to terminal), TCP optimization may also be applied to different elements along the network (Figure 8.1). The goal of such techniques is to adapt inherent TCP behavior in retransmissions and congestion control, by performing TCP tweaking, tunneling, splitting, acknowledgment (ACK) handling or any other technique acting on TCP mechanisms in the wireless domain. Special emphasis will be given to this group of techniques along the chapter.

Another issue to be considered when optimizing the service performance is the amount of data that is transferred from peer-to-peer. Due to the current bandwidth limitations over wireless links, the amount of data to be transferred through them should be reduced as much as possible. Different compression techniques (for protocol headers and application contents) may be applied to minimize the amount of data while maintaining a proper quality at application layer. In some cases, the removal of specific data not important for the end-user (e.g. HTTP banners, HTML comments, etc.) may also be performed at application level. Anyway, as new radio technologies evolve to support higher data rates, they will allow higher quality and richer contents to be delivered to the mobile terminals.

In addition to protocol and network optimization, the inclusion of new *ad hoc* elements like Performance Enhancing Proxies (PEP) or traffic shapers to the network architecture is being integrated within mobile and corporate networks with the aim to accommodate the protocols' behavior and content sizes to the wireless environments. These network elements may include a huge variety of features, such as protocol adaptation, content optimization, Internet e-mail acceleration, optimization of proxy features, etc.

Aforementioned techniques can be jointly applied in a proper way in order to maximize the end-user experience of cellular customers. Some results associated to the different optimization techniques will be also shown along the rest of this chapter.

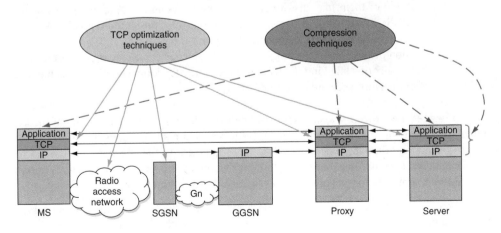

Figure 8.1 Location of optimization techniques

The remaining of the chapter is structured as follows. Section 8.2 describes the optimization methods at network level. Section 8.3 provides an overview of the transport optimization techniques. Section 8.4 outlines the compression techniques at different levels. Finally, section 8.5 introduces the role of the PEPs in the mobile networks.

8.2 Network-Level Optimization

A deep analysis of network-level optimization process might need a complete separate book, if we want to cover several radio technologies. Therefore, this section tries to summarize the main concepts to be taken into account when optimizing a network from the service performance point of view, without going into the details of each particular technology.

If we wonder what the final goal of the network optimization process is, we could think about three different dimensions, from a network operator's perspective.

1. *Optimize operational efficiency*: Automating most time-consuming tasks, performing parameter and dimensioning optimization automatically and regularly (minimizing the OPEX).
2. *Cost reduction*: Minimizing the CAPEX to support current or forecasted traffic, reusing as maximum the current infrastructure.
3. *Quality improvement*: Enhancing the network quality with minimum investments. Both, from the viewpoint of the network performance and also from the users.

This section is focused on the third dimension (i.e. quality improvement), which is the one that mostly affect the end-user performance. The first and second dimensions are more related to operator's perspective and would need a separate treatment, although some hints have already been introduced in Chapter 7.

As already described in Chapter 5, the main KPIs at network level that affect the service performance are throughput and latency. But few other network indicators are also important to optimize the quality, such as jitter (or delay variation), Bit Error Rate (BER) and outage time. Note that these indicators are some times related to each other in a way that the improvement of one of them may also improve the other. As an example, let us consider a decrease in the BER, which leads to less number of retransmissions at radio level and consequently to a lower jitter and higher effective throughput.

The optimization of the network is not an easy task, since every element or interface may be a quality bottleneck of some kind. However, some general recommendations to achieve an optimum network performance are proposed below.

- *Remove bottlenecks with proper end-to-end network dimensioning*: This is one of the key points to achieve an optimum performance. Both user and control planes must be properly dimensioned along the whole network, from radio interface to the core (see section 5.5 for further details). It is also recommended to consider forecasted traffic per cell for a certain time in future, not only for existing services but also for potential future services.

- *Transport network topology*: On many occasions, a user accessing to an Internet service through a mobile network is suffering from long delays, which are not completely under the control of the operator, unless some kind of Service Level Agreement is committed between operator and external service provider. However, when the network operator is providing access to proprietary services, it should take this aspect into account by, for example, minimizing the number of hops from the base station to the server. In general, an optimum transport network topology (avoiding very long paths) will minimize the network latency and maximize the user perception.

• *Cellular network topology*: In particular, the optimization of the Routing Area (RA) sizes is very important in (E)GPRS due to several factors: first, a mobile user suffer from long outage times moving to a new RA; secondly, the bigger the RA, the higher the signaling due to paging messages in that area (proportional to the number of cells per RA), but on the other hand, the smaller the RA, the higher the number of RA updates. Also location of RA borders is important to consider, as cell changes between different RA are far more costly than those taking place inside the same area.

• *Radio channel management*: Most radio technologies require specific procedures in charge of establishing a radio channel whenever some data need to be sent through the radio interface, and afterward, releasing the channel (or downgrading the bearer) when no more data is arriving. The delay associated to these procedures is especially harmful when traffic is bursty, since the frequency of such procedures is higher. Hence, the optimization of the radio channel management will also optimize the performance. For instance, (E)GPRS defines the Extended Uplink Temporary Block Flow (UL TBF) mode feature (defined in 3GPP release 4 [19]), which allows the extension of the UL TBF duration during a configurable period of time (from 0 to 5 seconds), even though the terminal does not have any UL data to transmit.[1] In WCDMA, optimal inactivity timers for Dedicated Channels (DCHs) carrying packet data and optimization of the conditions that govern the transition from FACH to DCH (based on the amount of Radio Link Control (RLC) data to be transmitted) are two of the optimization paths to be applied [8] (see also Chapter 5).

• *Quality of Service (QoS) management*: The support of QoS features is quite important if we want to prioritize particular data flows or guarantee certain delay requirements for particular services at the expense of degrading the quality of other non-critical services (in terms of delay or throughput). Those features become crucial for new real-time services like audio/video streaming, Push-to-Talk over Cellular, etc. An example of how basic QoS prioritization may affect user performance and network capacity can be found in section 6.5. However, in order to provide a QoS to real time and other high demanding users, QoS priorities are not sufficient, and other mechanisms as Admission Control and Quality Control are required to ensure that the system does not allow to start new data calls which may cause users already in the system to degrade their performance under required values.

• *Parameter settings*: Network parameters may have important impact on both delays and throughput user KPIs. However, it is not always possible to find an optimal configuration that provides the best performance for every service in any location inside the network. For example, maximizing the initial bit rate allocated to a connection might waste resources and increase the delay if the average quality in the cell is poor, but it could improve the

---

[1] A similar (E)GPRS procedure is available for Downlink (DL) from 3GPP Rel'97 called Delayed DL TBF release.

performance of cells with good quality. However, as a rule of thumb, it is much easier to increase the degradation by setting wrong parameter values than to get a small improvement by optimizing them. Therefore, parameter optimization must be done in combination with an exhaustive monitoring of the network KPIs, and also with active measurements that tracks the performance of the user. It has to be also taken into account that under certain circumstances (i.e. cells with low traffic), small variation in the parameters may cause short-term great variation in the performance indicators (even degradation) while the long-term tendency (with reliable data) might still mean an improvement. Applications which provide monitoring capabilities of the parameters and network KPIs through the whole network are needed in order to efficiently follow up parameter optimization activities.

- *Mobility management*: This group of procedures is responsible for providing a 'service continuation' while on the move. Some technologies like (E)GPRS are specially influenced by the mobility, in the sense that huge outage times degrade the service performance during cell reselections (from 1 to several seconds). Long outage times are probably the most harmful event for the TCP performance in moving environments and therefore for the upper application performance [9]. Although new features have been standardized for future releases of EGPRS (e.g. Network Assisted Cell Change (NACC) [20] to minimize cell-reselection duration (to around 500 ms), this is not enough to support conversational-type of services over (E)GPRS while on the move. Other technologies like WCDMA include soft handover functionalities, which do not suffer from outage times during normal cell changes. Only inter-system and inter-frequency handovers can create a pause in the communication (see also section 5.4).

There exist many other techniques that are applied in the network, though they are targeting to improve upper layer performance (e.g. TCP). The use of buffer congestion management methods or the incorporation of PEPs could be included in this group; these methods are described in detail in the following section.

8.3 Transport-Level Optimization

Cellular link characteristics differ a lot from wired links, which have low losses rate, short latencies and constant bit rates as main characteristics. Cellular networks make use of specific link layer protocols capable of providing a reliable bearer for the transmission in the air interface. However, radio link protocols are not able to hide air interface characteristics to higher layers which also have to be optimized for their usage in a cellular environment.

In that sense, new *ad hoc* protocols have been developed to be applied to wireless environments in order to optimize the end-user performance, as it is the case of Wireless Access Protocol (WAP). However, current Internet protocols (like TCP/IP) have also been used in the wireless architecture as an inherent way of merging mobile with fixed networks. Since Internet transport protocols were initially developed for wired networks, its use in wireless applications may lead to a non-optimum service performance.

TCP behavior is much more problematic than other existing transport protocols (like User Datagram Protocol, UDP), due to its congestion and flow control mechanisms. That is why TCP is subject to a significant research effort in recent years and now the support of TCP-based applications is becoming very common within mobile environments.

A first classification of TCP optimization techniques might be done according to:

- TCP improvements and recommendations proposed by the IETF [1].
- Buffer congestion management techniques.
- TCP optimization techniques within the network (e.g. in a proxy).

8.3.1 Standard TCP Recommendations from IETF

TCP improvements and recommendations for wireless links proposed by the IETF can be classified as safe for its use on the general Internet. In [1], different TCP optimizations for second and third generation wireless networks are presented. Main recommendations are:

- Appropriate window size
- Increased initial window
- Limited transmit
- TCP packet size
- Selective acknowledgment (SACK)
- TCP Timestamps option.

8.3.1.1 Appropriate Window Size

TCP over wireless links should be set within an appropriate receiver window size based on the Bandwidth Delay Product (BDP) available in the system. The receiver advertized window, in charge of the flow control from the receiver to the transmitter, has to be at least as large as the BDP; otherwise the receiver TCP layer will limit the maximum achievable bandwidth (section 5.4.1.2).

For example, let us consider a GPRS terminal using 3 TSL in downlink (DL) and 1 TSL in UL. Assuming a typical GPRS RTT around 1.1 seconds and effective RLC throughput of 30 kbps, BDP can be simply computed as:

$$BDP = Bandwidth * RTT = 30\,kbps * 1.1\,s = 4125\ bytes$$

For an EGPRS terminal with same TSL capability, RTT can be estimated as 800 ms and the bandwidth as 115 kbps. It entails to a BDP of 11.5 kB.

Some operating systems have not properly configured its TCP advertized window for such environments. For instance, Window NT 4.0[2] has a default advertized window of 8760 bytes. Therefore, if we were using a computer with a Window NT 4.0 operating system and aforementioned phones as wireless modem for Internet access, advertized window should not be a problem in case of GPRS but it would be limiting the available bandwidth in the case of EGPRS.

---

[2] Windows 98 had same default values, but already Windows 2000, Windows XP and different Linux versions implement higher advertized windows (i.e. 32 kB or 64 kB).

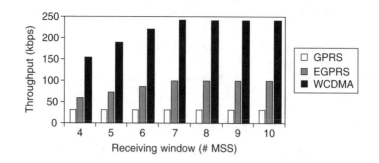

Figure 8.2   Impact of advertized window on FTP performance

Table 5.4 shows typical RTT and bandwidth values for different radio technologies, as well as their subsequent BDP values. From this table, we can compute recommended receiving window sizes to achieve an optimum performance at TCP layer.

As an example, FTP performance results for different advertized window configurations are shown in Figure 8.2. They assume an effective DL throughput at RLC level of 30 kbps (GPRS, 3 + 1 TSL), 115 kbps (EGPRS, 3 + 1 TSL) and 384 kbps (WCDMA).

As shown in the Figure 8.2, for EGPRS terminals the pipe is not fully utilized if advertized window is lower than 7 MSS, which matches with the theoretical estimations of the BDP. Although WCDMA even achieves higher bit rates than EGPRS, this threshold is similar to EGPRS (i.e. advertized window <7 MSS) due to the fact that the RTTs are smaller (around 200 ms). As expected, GPRS is not affected by the advertized window since this is always higher than the BDP.

Advertized window should be high enough to enable using all the available bandwidth but if advertized window is much higher than the BDP, it could lead to a worse performance due to buffer overflows and consequent TCP retransmissions. This effect is not visible in the figure, but an example of the impact can be seen in section 6.4.5.1. Therefore, advertized window should be set to a value slightly higher than the BDP in order to use the capacity efficiently and at the same time it do not impact on network congestion and loss recovery capabilities.

8.3.1.2 Increased Initial Window

As explained in Chapter 5, one of the main characteristics of the wireless links is the high latency compared with wired links. This fact makes much longer the slow start phase both at the beginning of the TCP connection, after a Retransmission Timeout (RTO) or when restarting the connection after a long idle period. By increasing the initial window, slow start phase will need less cycles to fill the pipe, providing a better performance.

Initially, 1 segment was the default initial window value in first specifications, although current applications tend to use 2 segments. In [2] an optional standard for TCP to increase the initial window is specified, and its upper bound is given by the formula:

$$IW = \min(4 \cdot MSS, \max(2 \cdot MSS, 4380 \text{ bytes}))$$

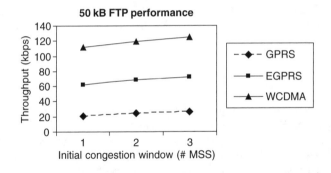

Figure 8.3   Impact of initial window on FTP performance

With this extension, a TCP transmitter might use up to 3 or 4 segments in the initial window (without exceeding 4380 bytes). The common case of a MSS of 1460 bytes (MTU of 1500 bytes, and no IP or TCP options including extra header), 4380 bytes allows the transmitter to use 3 segments for the initial window. Lower MSS imply the possibility of using even 4 segments in the initial window but would not improve the performance from the BDP utilization point of view. Aforementioned upper bound entails a change in the initial window limits set in previous IETF recommendations [3], where maximum initial window was set to 2 segments.

The scenario proposed in [2] for increasing initial window only includes the slow start at the beginning of the connection and optionally after a long idle period. But the proposal does not include the initial window after a RTO (loss packet) as the RTO is considered by TCP as a congestion indicator.

Figure 8.3 shows the performance of FTP downloads versus the initial congestion window (assumptions for each technology are the same as in Figure 8.2). File size used for the analysis is 50 kB. The smaller the file size the higher the performance improvement when increasing the initial congestion window. This is due to the longer proportion of time that the initial slow start phase takes when downloading small files. Although typical average sizes in FTP services might be slightly higher than 50 kB, these results could be extrapolated to other TCP-based services like Web browsing or WAP 2.0, which typically uses contents compounded of small objects. In those cases, an increased initial window might lead to a noticeable improvement of the overall performance.

8.3.1.3 Limited Transmit

Limited transmit [4] tries to improve the performance of Fast Retransmit/Fast Recovery algorithm in the case of small congestion windows. When there is small amount of outstanding data, i.e. unacknowledged data sent, it is difficult the fact that three more packets arrive to the receiver after a segment loss. Therefore, the receiver is not able to send enough number of duplicate ACKs (usually three) to the transmitter in order to trigger the Fast Retransmit/Fast Recovery algorithm.

With limited transmit, whenever a transmitter has unsent data waiting for transmission, a new segment will be sent upon the arrival of two duplicate ACKs. Thus, receiver could generate the third duplicate acknowledgment needed in the transmitter to trigger Fast

Retransmit/Fast Recovery algorithm, avoiding an RTO and the subsequent slow start phase. However, this option is not generally available in all systems, and would depend on the software implementation.

8.3.1.4 TCP Segment Size

The slow start needs a certain number of cycles in order to fill the available bandwidth, which depends on the BDP. Additionally, high latencies of wireless links make first cycles really critical for the end-user throughput. During the first cycles, the bandwidth utilization is very limited in some cases. In this respect, another variable to be considered is the size of the TCP segments. The larger the Maximum Segment Size (MSS) the shorter the slow start phase will take to fill the pipe, due to the fact that during slow start the increment of transmitted bytes is in units of segments. In case of bigger segments, the bandwidth utilization during first slow start cycles is higher and the BDP can be reached quicker. In addition, as the reader could guess, MSS also has an impact on the total packet overhead introduced by the different protocols.

Another drawback of using small TCP segment sizes is that the increase of the number of ACKs that are sent back to the transmitter. ACKs also consume available bandwidth in the reverse direction and, in some radio technologies like (E)GPRS, it has an extra impact on the performance as the number of TBF establishments and releases increase (due to the burstiness of the traffic). In this particular case, RLC signaling needed for the TBF establishments may consume resources in both directions (as described in section 5.3.1).

Since radio link layer offers retransmissions and a transparent fragmentation, higher packet sizes at network/transport layer do not imply higher probability of transmission failure. Then, a high MSS is beneficial even in the case of bad radio link conditions. Figure 8.4 shows the degradation of FTP throughput when decreasing the MSS. Only for services which require very low delays and fast error recovery, the utilization of small TCP packets could provide performance improvements.

8.3.1.5 Selective Acknowledgment

During a TCP connection, receiver includes in the ACK the sequence number of the last successfully received byte (i.e. last successfully received segment). This procedure is referred

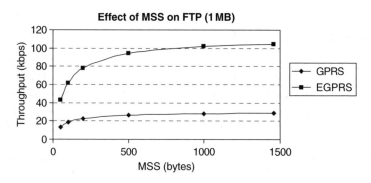

Figure 8.4   Impact of MSS on FTP throughput

to as cumulative ACK. When several packet losses occur, only cumulative ACK approach allows the transmitter to know the loss of the first packet.

SACK is an optional feature [5] that allows the receiver to indicate to the transmitter that all the segments have arrived successfully. That way, the transmitter can do a selective retransmission instead of retransmitting only the first lost packet and wait for the next ACK (one RTT) to receive the notification of new losses.

Hence, SACK feature is very beneficial in case of large BDP paths, such as wireless networks. In this kind of link, high latencies make much longer recovery time in case of several losses. However, the drawback of SACK would be an slight increase in the header overhead (up to 8 extra bytes).

There is one additional issue to consider: in order for SACK to be used, both the client and the server must support it and agree to its utilization during the establishment phase. Therefore, even if client is optimized with SACK for better wireless performance, if it tries to access to a server that does not support SACK, no improvement will be experienced by the user.

8.3.1.6 Timestamp Option

Standard TCP implementations make a RTT measurement per congestion window. It can lead to a poor estimation of the actual RTT, which is worst as congestion windows grow. Wireless link characteristics (high variability of the delay, delay spikes, resource sharing) entails to the need of an accurate RTT estimation in order to reduce the number of the spurious retransmissions.

With Timestamp Option [6], the sender can compute the RTT per segment, even for the retransmitted ones. The improvement in the RTT estimation allows TCP RTO to adapt quicker to sudden delay variations, leading to a better TCP performance and reduced probability of spurious retransmissions.

However, Timestamp option is implemented by adding 12 bytes in the TCP header (32 bytes in total), and also requires that both server and client support it. Despite this increment of the header, Timestamp has been demonstrated to be useful in cellular environment [8].

8.3.2 Buffer Congestion Management

TCP responds to congestion by closing down the window and invoking slow start. Long-delay networks take a particularly long time to recover from slow start, so it is really important to avoid congestion. To remedy this, active queue management techniques have been proposed as enhancements to be implemented inside wireless core networks. Buffer congestion is a risk especially in the DL buffers of the bottleneck mobile network elements (e.g. 2G-SGSN in (E)GPRS). Therefore, appropriate congestion control methods should be implemented.

The following sections show some buffer congestion management techniques that try to prevent the congestion collapse by controlling the amount of data queued in the different hops of the end-to-end path, and distributing the discarded packets among different users, to force fast retransmission recovery instead of RTOs and slow start.

8.3.2.1 Random Early Detection (RED)

The principle of RED algorithm is to start dropping arriving packets probabilistically depending on the estimated average queue size [10]. By discarding some packets before the queues are full, the senders will decrease its transmission rate as soon as packet losses are detected. That way, the congestion probability in the buffers will also decrease. RED algorithm increases the dropping probability of incoming packets as the buffer occupancy grows. If the buffer occupancy exceeds a threshold, all the arriving packets will be discarded.

With this method, 'global synchronization' effect is avoided. In global synchronization phenomenon, routers' load oscillates from empty to congest mostly due to TCP dynamic congestion control mechanism of different connections acting at the same time.

8.3.2.2 Fast TCP

Fast TCP algorithm tries to decrease the amount of incoming DL packets when DL buffers exceed a certain threshold (close to congestion) by delaying the UL ACKs (Figure 8.5a). Typically DL buffers get congested before UL buffers, since most of the services generate traffic mainly in the DL direction.

When UL ACKs are delayed, TCP transmitters wait a longer time before sliding its transmission window. This mechanism provides some extra time for the DL buffers to recover from congestion. At the same time, congestion window grows slower because of the delay of the ACKs.

However, the delays applied to the UL ACKs cannot be very high as it would probably imply increased RTO at the transmitters. Therefore, there has to be a trade-off in the applied delays in order to compensate the buffer overflow and the retransmission delay.

8.3.2.3 Window Pacing

The principle of Window Pacing is similar to Fast TCP, i.e. when downlink buffers exceed a predefined threshold, it tries to decrease the amount of stored data by modifying UL traffic characteristics. However, instead of delaying UL ACKs, Window Pacing decreases the advertized window in ACKs (Figure 8.5). That way, transmission rate is decreased as well. Figure 8.5b depicts the Window Pacing basis.

Window Pacing offers the advantages of the absence of dropping packets (as done in RED algorithm) and there is no risk to unintentionally force an RTO (as in the case of Fast TCP).

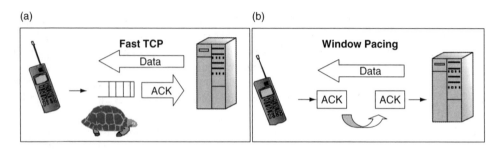

Figure 8.5   TCP ACK handling mechanisms

However, Window Pacing presents the drawback of a heavier processing task since TCP header has to be modified by the network element implementing it (i.e. SGSN). Additionally, this method is not feasible if IP security is used, as TCP headers would be protected.

8.3.3 TCP Optimization in an Intermediate Node

These optimizations include TCP changes specifically developed for wireless environments, most of them implemented within the PEP [7] (see also section 8.5). Such TCP modifications are usually non-standard and should be used carefully, understanding their consequences, especially toward the Internet and end-to-end requirements of TCP connections.

8.3.3.1 TCP ACK Handling

There are many optimization features based on ACK handling. A TCP implementation might combine some of them along with other kind of mechanisms.

- *Redundancy elimination*: An intermediate node monitors the latest TCP ACK sent over the wireless link from the receiver to the transmitter. The intermediate node detects any possible retransmitted packet, checking whether the ACK of this retransmitted segment has already arrived to it. If so, the retransmitted packet is discarded. This procedure increases the radio link efficiency and accelerates the slow start algorithm in case of spurious RTO. Figure 8.6 shows the principle of the redundancy elimination.

- *Local TCP ACK*: An intermediate node acknowledges locally the received TCP segments. With this mechanism slow start is speeded up avoiding typical long waiting delays during slow start phase in large BDP links. Note that this procedure may imply some end-to-end concerns, as packets are acknowledged which have not yet been delivered to their destination.

- *Local TCP retransmission*: Systems implementing local TCP ACKs are required in order to implement local TCP retransmission. The intermediate node retransmits (toward the receiver) lost segments on the air interface. In order to manage the retransmissions, the intermediate node has to implement mechanisms to determine when to retransmit lost data. This is fulfilled by storing the unacknowledged TCP segments in the intermediate node and the

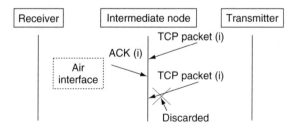

Figure 8.6   Redundancy elimination

usage of local retransmission timers. This method reduces the number of spurious retransmissions (and the subsequent slow start phase) in the transmitter.

• *Deletion of duplicate ACK*: Duplicate ACKs from the receiver to the transmitter are dropped in the intermediate node. Additionally, the intermediate node may perform local retransmissions based on the duplicate ACK information. This technique is an enhancement of the local TCP retransmission method and its goal is to avoid fast retransmit/fast recovery phase in the end transmitter.

Most of these solutions can, however, be considered as partial implementation of a TCP connection splitting, which consist of establishing an intermediate connection with a partner element inside the operator's network (proxy) that will connect with a different TCP connection to the remote server in the Internet. This solution is further described in the following sections.

8.3.3.2 RTT Adjustment

The delay spikes in the radio link as well as the variation of the TCP packet size usually produce sudden RTT variations which lead to spurious retransmissions. RTT adjustment tries to keep RTT variation under control in order to avoid RTOs. Aforementioned RTT variation can be controlled by means of a proper delay of the ACKs coming back from the receiver to the transmitter, thus decreasing the probability of RTO at the transmitter. Since different segment sizes may suffer different delays in the network, one possibility is to consider the TCP segment size as a variable to apply a proper delay to each segment.

8.3.3.3 Split TCP

End-to-end TCP connection is split into two separate connections: one from the server to a gateway inside the operator's network, and another from the gateway to the mobile station. By splitting the TCP connection, each of the resulting connections can be optimized for their special domain characteristics (e.g. wireless and wired). However, breaking the end-to-end connection might imply some negative consequences. Main benefits and drawbacks of split TCP for wireless links are summarized below.

Benefits

• *Shield problems*: The use of two separate connections is the best way to shield specific problems from one link to another; local solutions can be applied at each side. For instance, packet losses in wireless side can be treated as an extra delay due to transmission error in the radio link whereas in the wired connection it will be used as a congestion indicator. Potential negative consequences to the general Internet of TCP tweaking feature for wireless links are therefore avoided as they are going to be used only within the wireless environment.

• *Wireless link model*: An accurate model of the radio link entails to a better adaptation of the TCP to its specific characteristics. As these characteristics are well known, TCP can be adapted to them, improving its performance in wireless.

- *Deployment of wireless solutions*: Split connection allows an earlier deployment of proposals for wireless TCP. Only mobile devices and intermediate wireless nodes have to implement them. Hence, it is not needed that these new proposals become widely implemented in the general Internet, which usually takes quite a long time.

Drawbacks

- *Breaks end-to-end semantics*: One of the main problems of breaking end-to-end communications is that it disables the usage of the IP layer security mechanisms.

- *Extra delay*: New element in the path introduces an extra delay, even more when packets have to pass through the entire protocol stack and being processed by a new TCP-aware entity.

8.3.4 Datagram Congestion Control Protocol (DCCP)

UDP is commonly used by applications with strict time requirements and error tolerance. However, when the amount of UDP traffic increase in the network, there is an important risk of congestion. Since multimedia traffic is nowadays increasing day by day, different research areas have been created to solve such a problem by introducing some congestion control for datagram transmissions.

Recently, DCCP [17] was developed for time-sensitive applications, including a configurable congestion control mechanism. This new solution tries to merge some features of UDP and TCP into a single protocol, although new features are also supported. DCCP will provide a standard way to implement congestion control negotiation and enable the use of Explicit Congestion Notification (ECN) for applications. Main DCCP characteristics are listed below.

- A reliable handshake for connection set-up and tear-down. The negotiation of congestion control attributes is performed during connection set-up.
- *Buffer control*: Packets may be dropped from application's queue if they are old compared to other arriving data.
- *Unreliability*: DCCP does retransmit datagrams, although some protocol options are retransmitted as required to make feature negotiation and ACK information reliable.
- Congestion control incorporating ECN.
- *Two half connections (for UL and DL)*: Normally traffic in different directions has a different behavior; that way each connection may have its own congestion control parameters, both negotiated during connection set-up, and specified by a Congestion Control Identifier (CCID).

The CCID for a half-connection describes, for instance, how the sender limits data packet rates, the management of the congestion window, and how the receiver sends congestion feedback via ACKs. Each congestion control mechanism supported by DCCP is assigned a congestion control identifier, or CCID: a number from 0 to 255. Initially, TCP-like congestion control is defined by CCID 2, which is appropriate for flows that want to quickly take advantage of available bandwidth, while CCID 3 defines a TCP-Friendly Rate Control [18], which is appropriate for flows that require a steadier send rate.

There also exist different ACK formats. The CCID for a connection determines how much ACK information needs to be transmitted. In CCID 2 (TCP-like), this is about one ACK per 2 packets, and each ACK must declare exactly which packets were received (ACK Vector option); in CCID 3 (TFRC), it is about one ACK per RTT, and ACKs must declare at least just the lengths of recent loss intervals. DCCP packet is 12 bytes in length, although it is extensible for additional features.

Since DCCP is a congestion control protocol, it does not include any flow control mechanisms by means of receiving windows (as TCP does). It uses sequence numbers but they refer to packets instead of transmitted bytes.

A Data Dropped option in the protocol lets one endpoint declare whether a packet was dropped because of corruption, because of receive buffer overflow, etc. This facilitates the research into more appropriate rate-control responses for these non-network-congestion losses, although currently all losses will cause a congestion response.

8.4 Compression Techniques

Considering that the time it takes to transmit a given amount of data is equal to the size of the data divided by the expected throughput, one obvious way to enhance the end-user experience is to reduce the amount of data that needs to be transmitted. The techniques to do so are normally referred to as compression techniques, and we will in this section go through various techniques as well as analyse under what situations it would be beneficial to implement them.

The overall purpose of any type of compression is, as its name indicates, to reduce the amount of data that should be transmitted, especially through the most limiting links. By doing that, there are two immediate gains for the end-user.

1. *Time* – as it takes less time to transmit less amount of data.
2. *Money* – as the user normally pays for the amount of information he is transferring.

These two factors together make compression techniques an important method both now and in the foreseeable future. Typically, an operator would be able to implement compression techniques in intermediate agents inside its network, which would be applied to DL transmission (i.e. PEP), but it could also provide compression capability to their users by distributing the necessary software to be included in the subscriber's terminals or PCs (i.e. applicable to UL transmission). Additionally, from network's perspective, compression techniques lead to a higher capacity in terms of number of users, since the utilization of the resources is optimized.

8.4.1 General Fundamentals

Typically, compression is done just before the data is fed into the transmitter and decompression is done after the receiver has detected the signal. Compression may be generally understood as an end-to-end functionality.

Any compression technique aims to remove redundant information in the data that should be transmitted. It is important to understand the difference between the information a message contains and the actual way of coding it for the transmission. Consider the following example.

> *John wants to wish his friend Mary on her birthday, so he calls her up and says 'Happy Birthday Mary!' when she answers. Assuming that it takes John two seconds to say it, and that the telephone system uses PCM coding with 64 kbps, results in John needing 128 kB for sending his message.*
>
> *An alternative way could be for John to send an email. In this case his message is coded to ASCII text and he basically needs 1 byte per letter or 160 bits to send his message (for the sake of this example we neglect to include header information and addresses etc.).*

The example above might not be realistic, but it clearly illustrates that just by selecting the way how information is converted to bits, there are significant gains in terms of the amount of data that finally needs to be transmitted. All compression techniques aim to select the most cost efficient way of representing the information in the message to be transmitted.

Compression methods are typically divided into two main categories: *lossless* and *lossy* compression. The terms refer to whether any information is removed during the compression process so that the message cannot be exactly reproduced again.

Typically **lossy compression** can be applied when the receiver of the information is a human being, as the information loss normally causes such small distortion to the reproduced data so the human sense cannot notice it. **Lossless compression** on the other hand shall be used when there is a clear requirement that the information should not be changed whatsoever. This is, for example, of uttermost importance in communication between computers, or where a human receiver needs to notice all the details of the received message, for example in the transfer of a text file.

Another way to classify compression techniques is to divide them into *header compression* and *content compression*. These labels refer to whether the compression is done on the actual data content of the packet, or on the header of the packet. As the packet header is critical for routing the data to its destination, no information loss can be accepted, which means that for header compression, only lossless compression techniques can be used.

8.4.2 Content Compression Techniques

Content compression aims to reduce the amount of information that needs to be transmitted. As mentioned earlier, there are two main methods, lossless and lossy compression. In general, it can be said that the **lossless compression** is application independent, meaning that it does not assume anything about how the receiver will perceive the data. Instead, it could use a predefined algorithm which had been tested to work efficiently (or even optimal) for generic kind of data.

Lossy compression, on the other hand, is always application dependent, as it assumes the receiver do not need all of the original data, so a part of it can be removed. Typically, assumptions are that a certain distortion of image quality or music is not detectable by the human senses. Note that for wireless devices these assumptions can result in rather dramatic reductions of the amount of data sent, as the mobile device normally have a smaller screen than normal PCs or laptops, and the user normally have noise in the background.

Table 8.1 Content compression types: pros and cons

Compression type	Pros	Cons
Non-destructive	No information loss	Lower compression ratio
Destructive	Higher compression ratio; possible to trade compression ratio against amount of data	Information cannot be exactly reproduced; high compression ratio normally affects the receiver perception of the information

The method to be used for compressing data mainly depends on the receivers' ability to detect whether there is information loss or not. Table 8.1 summarizes the pros and cons of the different compression approaches.

Currently, there exists many content compression methods, some of them using a non-destructive approach such as Huffman coding or Lempel-Ziv (LZ), and other using a destructive approach like JPEG, MP3 or AMR.

• LZ compression methods are the most popular algorithms for lossless storage. Some variations of this technique are used in PNG and GIF images, as well as serve as the basis for the Zip method. LZ methods utilize a table-based compression model where table entries are substituted for redundant data.

• Huffman coding uses a specific method for choosing the representations for each symbol, resulting in a prefix code that expresses the most common characters in the shortest way possible. Huffman coding is the most effective compression method of this type.

• JPEG is a commonly used standard method for compressing photographic images. JPEG itself specifies only how an image is transformed into a stream of bytes, but not how those bytes are encapsulated in any particular storage medium. A further standard, created by the Independent JPEG Group, called JFIF (JPEG File Interchange Format) specifies how to produce a file suitable for computer storage and transmission (such as over the Internet) from a JPEG stream. In common usage, when one speaks of a 'JPEG file' one generally means a JFIF file, or sometimes an Exif JPEG file. There are, however, other JPEG-based file formats. JPEG/JFIF is the most common format used for storing and transmitting photographs on the World Wide Web. It is not well suited for line drawings and other textual or iconic graphics because its compression method performs badly on these types of images. (The PNG and GIF formats are in common use for that purpose. GIF having only 8 bits per pixel is not well suited for color photographs, but PNG may have as much or more detail than JPEG.) The MIME media type for JFIF is image/jpeg (defined in RFC 1341). An example of different image compression algorithm performance is shown in Figure 8.7.

• MP3 (i.e. MPEG-1/2 Audio Layer 3) is an audio compression algorithm capable of greatly reducing the amount of data required to reproduce audio, while sounding like a faithful reproduction of the original uncompressed audio to the listener. Many listeners accept the MP3 bit rate of 128 kbps as near enough to CD quality for them; this provides a compression ratio of approximately 11:1, although listening tests show that with a bit of practice, many listeners can reliably distinguish 128 kbps MP3s from CD originals. To some listeners,

Format	150×208 pixels	83×116 pixels	75×104 pixels
Bitmap (24 bits)	93.6 kB 	28.9 kB 	23.4 kB 
Bitmap (8 bits)	31.2 kB	9.6 kB	7.8 kB
GIF	14.6 kB	5.8 kB	4.8 kB
JPG	8.2 kB 	3.5 kB 	3.5 kB 

Figure 8.7   Content compression techniques

128 kbps is unacceptably low in quality. Quality of MP3 depends on quality of encoder and the difficulty in encoding the signal. Good encoders give acceptable quality at 128–160 kbps and very good quality is achieved at 160–192 kbps. Low-quality encoders never reach nearly transparent mode, not even at 320 kbps. So it is pointless to speak of 128-kbps or 192-kbps quality. A 128 kbps MP3 encoded with a good encoder might sound better than a 192-kbps MP3 file encoded with a bad encoder. An important feature of MP3 is that it is lossy – meaning that it removes information from the input in order to save space. As with most modern lossy encoders, MP3 algorithms work hard to ensure that the sounds it removes cannot be detected by human listeners, by modeling chacteristics of human hearing such as noise masking. The importance of this is that it can gain huge savings in storage space with appropriately reasonable and acceptable if not unnoticeable losses in fidelity.

• Adaptive Multi-Rate (AMR) is a lossy data compression scheme optimized for speech. AMR is adopted as the standard speech codec by 3GPP. The codec has 8 bit rates, 12.2, 10.2, 7.95, 7.40, 6.70, 5.90, 5.15 and 4.75 kbps (as depicted in Figure 8.8). The bitstream is based on frames which contain 160 samples and are 20-ms long. AMR enables better speech quality and increases the capacity of the network, but does not require any additional hardware investment. The solution relies on the capability of the AMR codec to adapt how effectively it performs according to the prevailing channel conditions. Unlike previous GSM speech codecs (FR, EFR and HR), which operate at a fixed rate and constant error protection level, the AMR speech codec adapts its error protection level to the local radio channel and traffic conditions. AMR actually comprises a family of codecs, which greatly increases its flexibility. In poor network conditions that produce a high amount of errors, more bits are used for error correction to obtain robust coding. However, when transmission conditions are good, fewer bits are needed for sufficient error protection and more can therefore be allocated for speech coding.

AMR is being adopted by many wireless applications, like Push-to-Talk over Cellular (PoC, described in section 3.7), which shall support AMR 5.15 as mandatory and default codec.

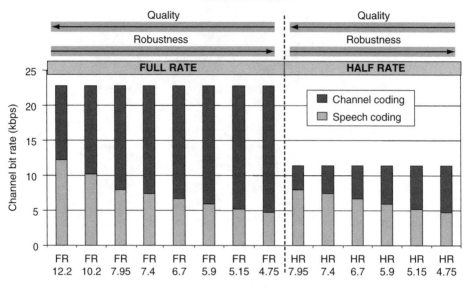

Figure 8.8   AMR codec modes

8.4.3 Wireless Specific Considerations

So far we have treated compression from a general perspective. But it must be noted that as it is done on a high level in the protocol stack, applications do not normally consider whether the information is transmitted over a wireless link or not. There are, however, some specific issues with wireless that are worth mentioning here. Wireless links are characterized by:

- Long RTT
- Highly error-prone
- Large overhead
- Limited radio resources.

These characteristics must be taken into account when developing and applying particular compression solutions. As indicated earlier, compression is done before the data is fed into the transmitter and decompression is done after the signal has been detected by the receiver. This is a general statement, and there is a lot of freedom on how to implement this. Two extreme cases are shown in Figure 8.9.

In alternative 1, we have placed the compression and decompression in the mobile device and application sever. In alternative 2, we have moved the compression functionality from the application server to a free standing box before the radio interface. Some times this box is just an independent network element close to GGSN (see section 8.5) or it could be part of the functionality of any existing network elements before the main bottleneck of the wireless network, i.e. the radio interface.

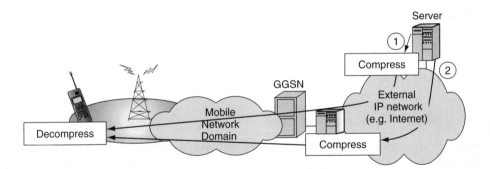

Figure 8.9   Compression/decompression process

With the compression functionality in the endpoints (mobile device and application server as shown in alternative 1) we can argue the following.

- The in-between network elements are not aware whether the data is compressed, which means that the dimensioning of the network is completely independent of the processing needed for compression.
- We are minimizing the transmission of information on all links, not only in the radio network.
- If compression is integrated in the application, large gains can be achieved as on application level the designer can make decisions on what information that needs to be transmitted at all.

By moving compression functionality closer to the GGSN (as in alternative 2) the following must be taken into consideration.

- The operator can implement compression on the last mile link (which is likely the bottleneck) without commitments from the service provider (or the one that runs the application server).
- This alternative is less scalable than alternative 1. By that we mean that if traffic grows heavily, the operator has to dimension his compression functionality properly. Note that it can become a significant effort for any network element to deal with the compression and decompression of all the data traffic carried through a GGSN.

Finally, for both proposed alternatives, there are some common issues as well.

- Compression is in general a processing intense process, which makes it a challenge to perform in a mobile device with limitations on processing and power consumption.
- The mobile device has limitations in screen size, resolution and color depth. It is also normally operated in a noisy environment. These factors affect the end-users possibility to perceive information losses during compression. Note, however, that the mobile network do not always have the knowledge if the information should be presented on the mobile device or on, for example, a laptop that is connected to the mobile.

- If the user makes use of secure communication, there are serious limitations on what compression techniques that can be applied and where. This is due to the fact that the data is encrypted, which results in a random sequence of bits, which will be difficult to compress efficiently. However, even in this case compression in the end points can still be fully applied if it is done to the data before secure communication is applied.

8.4.3.1 Header Compression in Wireless

Several IETF Working Groups are responsible for developing header compression schemes that perform well over links with high error rates and long link RTTs (like GPRS, WCDMA and CDMA2000).

First IETF recommendation for compressing TCP/IP headers for low-speed serial was developed by Jacobson [11]. Afterward, Degermark proposed an IP header compression (IPHC, both UDP/IP and TCP/IP) [12] which was standardized by the 3GPP for its use at PDCP layer for WCDMA and SNDCP layer for GPRS.

As an example, let us consider a voice communication that uses RTP/UDP/IP protocols. Assuming a typical payload at application layer of 20 bytes, the total overhead without any compression would be around 200% for IPv4 and 300% for IPv6. Van Jacobson method is able to compress the TCP/IPv4 header from 40 to 4 bytes (as an average) while Degermark's was extended to TCP/IP and UDP/IP (but not RTP), providing average compression values around 8 bytes for TCP/IPv4 and 5 bytes for UDP/IPv4.

The most advanced header compression technique is known as Robust Header Compression (ROHC) [13]. This technique has been recently included in 3GPP specifications. The main objective for ROHC has been robust compression of IP/UDP/RTP, but it also includes new solutions for TCP/IP. ROHC is intended to support IPv4, IPv6, Mobile IP, IPsec and TCP (including advanced TCP options) and even provides higher compression rates than previous schemes under the same conditions.

8.5 Performance Enhancing Proxies

A PEP is a new entity, added to the wireless operator's network, in charge of improving the end-user performance. Generally, wireless-PEPs (hereinafter referred to as PEP) aim at monitoring and accommodating the traffic going to/from the wireless interface from/to Internet. This adaptation is required due to the considerable mismatch between wire-line and wireless environments' characteristics.

PEPs may operate at any protocol layer, although typically they are commonly used to modify TCP and application layers' behavior [14]. These kinds of PEP are transparent to the network under which the mobile terminal is connected.

8.5.1 Transport Layer Features

Most of the transport layer's PEPs interact with TCP, since this is the most problematic transport protocol over wireless. Supported methods are generally oriented to overcome the inherent problems of TCP behavior, like TCP slow start. In that sense, different solutions may be applied, such as the following.

- *Adjust TCP parameters*: The goal is to set the optimum TCP parameters under the wireless environment (e.g. modify the MTU or receiving window from TCP ACKs) in order to improve the end-user throughput. An initial window of one or two segments may be harmful for short data transactions over wireless links. Therefore, it is quite important to increase the initial congestion window to 3 or 4 segments in order to improve the performance. An optimum TCP MSS is also quite relevant when considering the probability of packet losses in the wireless network. Generally, a large MSS[3] of about 1460 bytes performs optimally over cellular networks assuming that low layers are reliable (section 8.3).

- *ACK handling*: Many different methods related to ACK handling may be applied to enhance TCP performance. For instance, PEPs may perform local ACKs, local retransmissions or delete duplicate ACKs. Local ACKs allow improving the throughput in an environment with a large BDP, reducing the effect of TCP slow start. Local retransmissions of data segments lost on the path between the TCP PEP and the receiving end system may also improve the end-to-end performance. When a PEP receives a duplicated ACK in their way from the TCP receiver back to the TCP sender while detecting that the lost packet can be retransmitted to the TCP receiver, it may also delete the ACK (as described in section 8.3.3.1).

- *RTT adjustment*: PEP may perform continuous RTT measurements and analysis in order to keep RTT variation under control to avoid spurious retransmissions. For instance, a PEP might delay some ACKs depending on the size of the corresponding data segment or react to decrease bandwidth.

- *Connection multiplexing*: Although transport layer PEPs do not modify any application data, it can manage the TCP connections (requested from the application) in a more efficient manner, e.g. minimizing the number of TCP connections to be established.

- *TCP tunneling*: This approach is based on the idea of encapsulating TCP connection in another protocol (like UDP). However, it requires some changes at the other side of the connection.

Some of these transport optimization techniques were further detailed in section 8.3.

8.5.2 Application Layer Features

Application layer PEPs are aimed at improving the application protocol, minimizing the data to be transferred at application layer as well as optimizing the usage of transport layer by a particular application. They were initially used as a regular application proxy (for Web or mail) although lately they are including wireless-specific optimization features in order to improve the performance of different applications. Those features are especially relevant under networks with low bandwidth and long delays.

Application layer optimization techniques can be very powerful and in some cases almost mandatory in order to allow acceptable end-user experience. Mobile office type of application

---

[3] Limitation of the size of MSS comes from the MTU of the network. The sum, MSS + TCP/IP header size should in any case be smaller than MTU, in order to avoid further degradation by IP packet fragmentation.

transactions can be for instance greatly optimized for wireless allowing much better perform-ances. These techniques can be sorted into the following categories.

- *Content compression by applying standard or non-standard data compression schemes* (described in section 8.4.2)
- *Content and image optimization*

    - Filtering: Content that is not very relevant to the end-user and increases download times can be removed from Web pages (e.g. labels, digital signatures, HTML comments, large animated GIFs, etc.).
    - Downgrading: Image resolution is reduced before it is sent across the wireless link. This feature is especially useful over low-bandwidth devices or low-resolution devices, decreasing considerably the download times.

- *Server caching*: When a client has downloaded a particular document, and it is still in the cache, upon reload the server is asked to check the newest document available from the origin server against the version cached by the client. For many documents the differences between versions are small and it is enough to transmit only these small differences to reconstruct the newer version.
- *Business application support*

    - Reduces the signaling for, for example, email application.
    - Support for other applications including folder sharing.

8.5.3 PEP Integration in Cellular Networks

PEPs may be located in different places along the network. Generally, mobile network operators locate them just behind the network gateway (e.g. GGSN), although other locations within the mobile network domain are also possible. However, since PEPs are monitoring and modifying L4–L7 layers, this option is not compatible with Virtual Private Networks (VPN) when accessing to corporate networks.

Note that PEP functionalities might be even added to existing network elements (e.g. SGSN, BSC, etc.) without the need of including additional network elements. This option is quite efficient for some optimization methods, like for example those related to the buffer congestion management, which may be implemented in gateways, routers or radio access network elements. But in this case it would be the manufacturer of the network equipment who would be in charge of implementing the different algorithms, while in the other case, the hardware and software could also be provided by third parties.

Then, a PEP as separate network element is generally located behind the GGSN and before the Internet backbone, or inside Intranet corporate premises, depending on whether it is controlled by the mobile network operator or by the enterprise network. Figure 8.10 presents a scenario where the PEP is inside the mobile network operator's domain.

However, the location of PEP has some security implications. When a PEP is located in an intermediate node, the end-to-end security is lost, since IPsec is employed end to end, by encrypting the TCP header and application data, which is unintelligible to the PEP. In that situation, the operator must choose between using PEPs and using IPsec. One possibility is to split the 'security tunnels' into two parts, although this solution implies that the PEP should

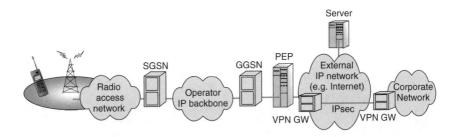

Figure 8.10   PEP located behind GGSN

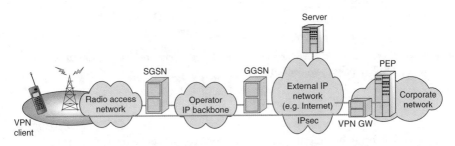

Figure 8.11   PEP located in corporate network

decrypt/encrypt the data, breaking the end-to-end security approach. Another solution is to allow the user to select which traffic must use IPsec or PEP processing.

There is some research in the area of simultaneous support of IPsec and PEP, although due to the complexity of IPsec mechanisms, it will take some time until any strong solution is standardized. One of the most interesting investigations in this respect proposes some changes in the implementation of IPsec in order to provide multi-layer IP security, i.e. by encrypting TCP headers and TCP payload [16] separately.

Some enterprises may want to ensure end-to-end security for their transactions. In this case, the PEP can be hosted in their network as illustrated in Figure 8.11. IP security is then used from the server in the corporate network to the mobile terminal. PEP clients are added to the mobile terminals (e.g. laptops) allowing increased optimization capabilities. For example, wireless-profiled TCP or content compression may require a PEP client to be applicable.

Finally, it should be noted that optimization do not always require a PEP as separate physical entity, but can as well be implemented in standard mobile network elements. Actually, some of these methods are more efficient near the radio interface (e.g. congestion methods, redundancy elimination) and should preferably be placed there. Furthermore, reducing the number of network elements also reduces the RTT and network management complexity.

8.5.4 Performance Improvement

This section provides some performance figures representing the improvement when introducing PEPs in mobile networks [15]. Final performance improvement is quite dependent on the type of service, contents and features supported by the PEP.

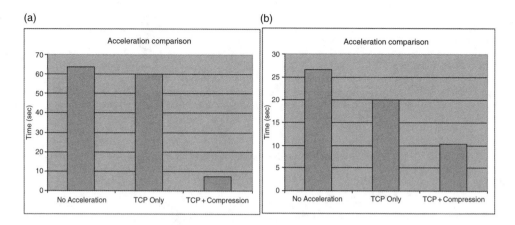

Figure 8.12 Comparison between application-level and transport-level optimization; (a) 400-kB image; (b) 100-kB Web page, 8 embedded objects [16]

Generally, services using large content sizes and low number of TCP connections (as for FTP sessions) will be highly improved by application layer optimization (e.g. compression) rather than TCP optimization. However, it also depends on how much the content is compressible.

An example of the improvement achieved by application-level and transport-level optimization techniques for a large object (400-kB image) and a Web page (100 kB) is shown in Figure 8.12.

Since the Web page has multiple objects and it is hosted in several different domain names, the overhead of DNS lookups and TCP connection set-up is much higher. Using the previously mentioned transport level optimizations, the speed up obtained equals 30% instead 6% obtained with a single large object.

Regarding application-level optimizations, they provide a compression factor over 800% for the single object. However, the improvement for the Web page is not so high since the objects were quite small and not very compressible. The actual speed up obtained when compressing text and images on the tested Web page was close to 250%. Still this factor is much higher number than the 30% provided by the transport-level optimizations alone. Therefore, application-level optimizations amount for a larger portion of the overall latency reduction, although their relative impact highly depends on the type of pages and their complexity.

References

[1] R. Ludwig, A. Gurtov and F. Khafizov, 'TCP over Second (2.5G) and Third (3G) Generation Wireless Networks', RFC 3481, February 2003.
[2] M. Allman, S. Floyd and C. Partridge, 'Increasing TCP's Initial Window', RFC 3390, October 2002.
[3] M. Allman, V. Paxson and W. Stevens, 'TCP Congestion Control', RFC 2581, April 1999.
[4] M. Allman, H. Balakrishnan and S. Floyd, 'Enhancing TCP's Loss Recovery Using Limited Transmit', RFC 3042, January 2001.
[5] M. Mathis, J. Mahdavi, S. Floyd and A. Romanow, 'TCP Selective Acknowledgment Options', RFC 2018, October 1996.
[6] V. Jacobson and R. Braden, 'TCP Extensions for High Performance', RFC 1323, May 1992.

[7] J. Border, M. Kojo, J. Griner, G. Montenegro and Z. Shelby, 'Performance Enhancing Proxies Intended to Mitigate Link-Related Degradations', RFC 3135, June 2001.

[8] G. Gómez, J. Torreblanca and J. Ramiro-Moreno, 'Assessing and Optimizing End User Performance over 2G/3G Cellular Networks', IASTED CSN'2004, Marbella (Spain).

[9] J. Torreblanca, G. Gómez and R. Cuny, 'Impact of RTT on TCP Performance over (E)GPRS Mobile Networks', IASTED CSN'2004, Marbella (Spain).

[10] B. Braden *et al.*, 'Recommendations on Queue Management and Congestion Avoidance in the Internet', RFC 2309, April 1998.

[11] Van Jacobson, 'Compressing TCP/IP Headers for Low-Speed Serial Links', RFC 1144, February 1990.

[12] Mikael Degermark, Bjorn Nordgren and Stephen Pink, 'IP Header Compression', RFC 2507, February 1999.

[13] Carsten Bormann *et al.*, 'Robust Header Compression (ROHC)', RFC 3095, July 2001.

[14] J. Border, M. Kojo, J. Griner, G. Montenegro and Z. Shelby, 'Performance Enhancing Proxies Intended to Mitigate Link-Related', RFC 3135, June 2001.

[15] Pablo Rodriguez and Vitali Fridman, 'Performance of PEPs in Cellular Wireless Networks', Microsoft Research, Cambridge, UK.

[16] Y. Zhang and B. Singh, 'A Multi-Layer IPsec Protocol,' Proceedings of 9th USENIX Security Symposium, Denver, Colorado, August 2000, Available at http://www.wins.hrl.com/people/ygz/papers/usenix00.html.

[17] E. Kohler, M. Handley and S. Floyd, 'Datagram Congestion Control Protocol (DCCP)', Internet-Draft, draft-ietf-dccp-spec-07.txt, 18 July 2004.

[18] I. Wu and T. Eckert, 'Router-port Group Management Protocol (RGMP)', RFC 3488, February 2003.

[19] 3GPP TS 44.060, 'Radio Link Control/Medium Access Control (RLC/MAC) protocol (Release 4)', V4.15.0, February 2004.

[20] 3GPP TS 44.060, 'Radio Link Control/Medium Access Control (RLC/MAC) Protocol (Release 4)', V4.16.0, May 2004.

Glossary

1xEV-DO	1X Evolution Data Optimized
1xEV-DV	1X Evolution Data and Voice
2G	Second Generation (Mobile Telephony)
3G	Third Generation (Mobile Telephony)
3GPP	Third Generation Partnership Project
3GPP2	Third Generation Partnership Project 2
4G	Fourth Generation (Wireless)
AC	Admission Control
ACK	Acknowledgment
ADSL	Asymmetrical Digital Subscriber Line
AF	Assured Forwarding
AMR	Adaptive Multi-Rate
APN	Access Point Name
ARMA	Autoregressive Moving Average
ARPANET	Advanced Research Project Agency Net
ARPU	Average Revenue Per User
ARQ	Automatic Repeat reQuest
AWND/awnd	Advertised Window
BCCH	Broadcast Common Control Channel
BCH	Broadcast Channel
BDP	Bandwidth Delay Product
BE	Best Effort
BER	Bit Error Rate
BLER	Block Error Rate
BSC	Base Station Controller
BSS	Base Station Subsystem
BTS	Base Transceiver Station
BW	Bandwidth
CAPEX	Capital Expenditure
CBQ	Class-Based Queuing
CCCH	Common Control Channel
CCID	Congestion Control Identifier
CDMA	Code Division Multiple Access
CDR	Charging Data Record
CEI	Customer Experience Indicators
CEM	Customer Experience Management

End-to-End Quality of Service over Cellular Networks: Data Services Performance and Optimization in 2G/3G
Edited by G. Gómez and R. Sánchez   © 2005 John Wiley & Sons, Ltd

CIR	Committed Info Rate
CLPC	Closed Loop Power Control
CN	Core Network
COPS	Common Open Policy Service
CORBA	Common Object Request Broker Architecture
CPCH	Common Packet Channel
CPU	Central Processing Unit
CQI	Channel Quality Indicator
CS	Circuit Switched
CSCF	Call Serving Control Function
CSI	Channel State Information
CWND/cwnd	Congestion Window
DCCP	Datagram Congestion Control Protocol
DCH	Dedicated Channel
DECT	Digital Enhanced Cordless Telecommunications
DiffServ	Differentiated Services
DL	Downlink
DNS	Domain Name Service
DPCCH	Dedicated Physical Control Channel
DPCH	Dedicated Physical Channel
DPDCH	Dedicated Physical Data Channel
DSCH	Downlink Shared Channel
DSCP	DiffServ CodePoint
DTX	Discontinuous Transmission
DVB-T	Digital Video Broadcasting Terrestrial
EDGE	Enhanced Data for Global Evolution
EDGE	Enhanced Data rates for GSM Evolution
EF	Expedited Forwarding
EFL	Effective Frequency Load
EGPRS	Enhanced General Packet Radio Service
ETSI	European Telecommunications Standards Institute
FACH	Forward Access Channel
FARIMA	Fractional Integrated ARMA
FDD	Frequency Division Duplex
FDDI	Fiber Distributed Data Interface
FDMA	Frequency Division Multiple Access
FER	Frame Erasure Rate
FER	Frame Error Rate
FER	Frame Error Ratio
FFT	Fast Fourier Transform
FGN	Fractional Gaussian Noise
FRBC	Frame Relay Bearer Channels
FTP	File Transfer Protocol
GERAN	GSM/EDGE Radio Access Networks
GGSN	Gateway GPRS Support Node
GLMS	Group and List Management Server
GMLC	Gateway Mobile Location Centre
GPRS	General Packet Radio Service
GPS	Global Positioning System
GRE	Generic Routing Encapsulation
GSM	Global System for Mobile Communication
GSN	GPRS Support Node
GTP	GPRS Tunnelling Protocol
HC	Handover Control
HIPERLAN	High Performance Radio Local Area Network

HIPERMAN	High Performance Radio Metropolitan Area Network
HLR	Home Location Register
HSCSD	High Speed Circuit Switched Data
HSDPA	High Speed Downlink Packet Access
HS-DPCCH	High Speed Dedicated Physical Control Channel
HS-DSCH	High Speed Downlink Shared Channel
HS-PDSCH	High Speed Physical Downlink Shared Channel
HS-SCCH	High Speed Shared Control Channel
HTML	HyperText Markup Language
HTTP	HyperText Transfer Protocol
HW	Hardware
ICI	InterCarrier Interference
ICMP	Internet Control Message Protocols
IEEE	Institute of Electrical and Electronics Engineers
IETF	Internet Engineering Task Force
IFFT	Inverse Fast Fourier Transform
IMS	IP Multimedia Subsystem
IMSI	International Mobile Subscriber Identity
IntServ	Integrated Services
IP	Internet Protocol
IR	Incremental Redundancy
IRP	Integration Reference Point
ISI	InterSymbol Interference
IW	Initial Congestion Window
KPI	Key Performance Indicator
KQI	Key Quality Indicator
LA	Link Adaptation
LC	Load Control
LDAP	Lightweight Directory Access Protocol
LDP	Label Distribution Protocol
LLC	Logical Link Control
LRD	Long Range Dependencies
LSP	Label Switching Path
LSR	Label Switching Router
MAC	Medium Access Control
MBWA	Mobile Broadband Wireless Access Network
MCS	Modulation and Coding Scheme
MIB	Management Information Base
MIMO	Multiple Input Multiple Output
MM	Multimedia Message
MMS	Multimedia Messaging Service
MMSC	MMS Center
MMSE	MMS Environment
MMSNA	Multimedia Messaging Service Network Architecture
MOS	Mean Opinion Score
MPLS	MultiProtocol Label Switching
MS	Mobile Station
MSC	Mobile Switching Centre
MSS	Maximum Segment Size
MT	Mobile Terminal
MTU	Maximum Transfer Unit
NACC	Network Assisted Cell Change
NCCR	Network Controlled Cell Reselection
NMS	Network Management System
NRT	Non-Real-Time

OFDM	Orthogonal Frequency Division Multiplexing
OFDMA	OFDM Access
OLPC	Outer Loop Power Control
OMA	Open Mobile Alliance
OSI	Open System Interconnection
OSS	Operations Support System
OVSF	Orthogonal Variable Spreading Factor
PACCH	Packet Associated Control Channel
PBCCH	Packet Broadcast Common Control Channel
PCCCH	Packet Common Control Channel
PCF	Packet Control Function
PCH	Paging Channel
PCU	Packet Control Unit
PDF	Policy Decision Function
PDN	Packet Data Network
PDP	Packet Data Protocol
PDP	Policy Decision Point
PDSN	Packet Data Serving Node
PDTCH	Packet Data Traffic Channel
PDU	Packet Data Unit
PEP	Performance Enhancing Proxy
PEP	Policy Enforcement Point
PFC	Packet Flow Context
PHB	Per-Hop-Behaviour
PIB	Policy Information Base
PLC	Power Line Communication
PoC	Push-to-Talk over Cellular
PPP	Point-to-Point Protocol
PS	Packet Scheduler
PS	Packet Switched
PTT	Push To Talk
QAM	Quadrature Amplitude Modulation
QoE	Quality of Experience
QoS	Quality of Service
QPSK	Quadrature Phase Shift Keying
RA	Routing Area
RAB	Radio Access Bearer
RACH	Random Access Channel
RAN	Radio Access Network
RAU	Routing Area Update
RED	Random Early Detection
RF	Reduction Factor
RFC	Request For Comments
RLC	Radio Link Control
RLP	Radio Link Protocol
RM	Resource Manager
RNC	Radio Network Controller
RNS	Radio Network Subsystem
RRC	Radio Resource Control
RRM	Radio Resource Management
RSVP	Resource ReserVation Protocol
RTCP	Real Time Control Protocol
RTO	Retransmission Timeout
RTP	Real-Time Transport Protocol
RTS	Right To Speak

RTSP	Real-Time Streaming Protocol
RTT	Round-Trip Time
RWIN	Receiving Window
SACK	Selective Acknowledgment
SAP	Service Access Point
SC	Single Carrier
SC	SMS Centre
SDCCH	Slow Dedicated Common Control Channel
SDH	Synchronous Digital Hierarchy
SDP	Session Description Protocol
SDR	Software Defined Radio
SDU	Service Data Unit
SF	Spreading Factor
SGSN	Serving GPRS Support Node
SINR	Signal-to-Interference-plus-Noise-Ratio
SIP	Session Initiation Protocol
SLA	Service Level Agreement
SM	Short Message
SME	Small and Medium Enterprise
SMS	Short Message Service
SMSC	SMS Centre
SMS-GMSC	Gateway MSC for Short Message Service
SMS-IWMSC	Interworking MSC for Short Message Service
SNDCP	Sub Network Dependent Convergence Protocol
SNMP	Simple Network Management Protocol
SNR	Signal-to-Noise Ratio
SOHO	Small Office/Home Office
SQM	Service Quality Management
SRD	Short Range Dependencies
SSI	Service State Information
STC	Space-Time Coding
STS	Start To Speak
SW	Software
TBF	Temporary Block Flow
TCP	Transport Control Protocol
TCP	Transmission Control Protocol
TDD	Time Division Duplex
TDMA	Time Division Multiple Access
TE	Terminal Equipment
TFT	Traffic Flow Template
TMF	TeleManagement Forum
TMN	Telecommunications Management Network
TSL	Timeslot
TTI	Transmission Time Interval
UCS	Universal Character Set
UDP	User Datagram Protocol
UE	User Equipment
UL	Uplink
UMTS	Universal Mobile Telecommunications System
URI	Uniform Resource Identifier
URL	Uniform Resource Locator
UTRA	UMTS Terrestrial Radio Access
VDT	Voice Delay Time
VLR	Visitor Location Register
VMSC	Visited Mobile Switching Centre

VoIP	Voice over IP
VPN	Virtual Private Network
WAE	Wireless Application Environment
WAP	Wireless Application Protocol
WCDMA	Wideband CDMA
WDP	Wireless Datagram Protocol
Web	World Wide Web
WLAN	Wireless Local Area Network
WMAN	Wireless Metropolitan Area Network
WML	Wireless Markup Language
WRR	Weighted Round Robin
WSL	Wireless Session Protocol
WTLS	Wireless Transport Layer Security
WTP	Wireless Transaction Protocol

Index

End-to-End Quality of Service over Cellular Networks: Data Services Performance and Optimization in 2G/3G
Edited by G. Gómez and R. Sánchez   © 2005 John Wiley & Sons, Ltd